APPLIED URBAN ANALYSIS

APPLIED URBAN ANALYSIS

A Critique and Synthesis

IAN CULLEN

Routledge
Taylor & Francis Group

LONDON AND NEW YORK

First published in 1984

This edition published in 2007
Routledge
2 Park Square, Milton Park, Abingdon, Oxon, OX14 4RN

Routledge is an imprint of Taylor & Francis Group, an informa business

Transferred to Digital Printing 2007

British Library Cataloguing in Publication Data
A CIP catalogue record for this book
is available from the British Library

Applied Urban Analysis
ISBN10: 0-415-41770-8 (volume)
ISBN10: 0-415-41934-4 (subset)
ISBN10: 0-415-41318-4 (set)

ISBN13: 978-0-415-41770-9 (volume)
ISBN13: 978-0-415-41934-5 (subset)
ISBN13: 978-0-415-41318-3 (set)

Routledge Library Editions: The City

Applied Urban Analysis

A critique and synthesis

Ian Cullen

Methuen

London and New York

First published in 1984 by
Methuen & Co. Ltd
11 New Fetter Lane, London EC4P 4EE

Published in the USA by
Methuen & Co.
in association with Methuen, Inc.
733 Third Avenue, New York, NY 10017

Phototypeset by Saxon Printing Ltd, Derby
Printed in Great Britain by Richard Clay (the Chaucer Press), Bungay, Suffolk

British Library Cataloguing in Publication Data
Cullen, Ian
  Applied urban analysis.
  1. Cities and towns
  I. Title
  910'.091732   GF125

  ISBN 0 416 36430 6
  ISBN 0 416 36440 3 Pbk

Library of Congress Cataloging in Publication Data
Cullen, I. G. (Ian G.)
  Applied urban analysis.

  Bibliography: p.
  Includes index.
  1. City planning—Methodology.   2. Anthropogeography—
Methodology. I. Title.
HT166.C83   1984   307'.12   84—20575
ISBN 0 416 36430 6
ISBN 0 416 36440 3 (pbk.)

Contents

List of figures   vi

Acknowledgements   vii
1 Introduction   1
2 The scientific method in planning
   and human geography   11
3 The economic and demographic bases   40
4 Spatial form   78
5 Perceptions, values and lifestyles   109
6 Structured locational decisions   139
7 Prediction, explanation and interpretation
   in applied urban analysis   173
8 Investigation and action   192
References   201
Name index   211
Subject index   215

Figures

2.1 The rational decision process   24
2.2 Policy evaluation given alternative
     optimality criteria   32
3.1 Integrated aggregate analysis   73
3.2 Integrated aggregate analysis incorporating
     the state   74
4.1 Spatially specific integrated aggregate analysis   105
5.1 The prisms of an individual's twenty-four-hour
     activity programme   129
5.2 The revealed preference and time geography
     models of human behaviour   133
5.3 Reflexive versions of the revealed preference
     and time geography models of human behaviour   135
5.4 Reflexive adaptation model of
     day-to-day lifestyles   136
7.1 An integration of dialectical and sympathetic
     approaches to the dynamics of choice and change   187

Acknowledgements

Many people have helped me to develop the ideas that are expressed in this book. These include Peter Cowan, Peter Willmott, Vida Godson, Elizabeth Phelps, Sandy Hammond, Erica Haimes and research students too many to mention by name. To each of these I should like to extend my thanks. I owe a further debt of gratitude to many people living and working in London who have, over the past ten years, so generously given me their time, patience and trust when I have called at their homes with a clipboard, a pencil and too many questions. Though this is not the report of an empirical study in the strict sense of that term it could not have been written but for the conversations I have had in the course of each of the empirical exercises in which I have been involved. I am most grateful also to Irene Morrish and to my mother, each of whom typed a draft of the book.

Finally I must reserve my warmest thanks for Lesley, whose support and encouragement have been unfailing.

1
Introduction

It is the aim of this work to contribute something to the advancement of research practice in planning and in human geography. In general such a contribution may be made either by precept or by example. The second approach at least has the merit of demonstrating its practicality as it assembles an argument. However, there comes a time when the scope afforded by this approach is no longer adequate. This is particularly the case when an area of research becomes the subject of a debate which goes right to the epistemological core of the academic disciplines upon which it is built.

Such is the case with research in planning and human geography at the present time. It has been and still is healthy in the sense that as an area of study it is popular. A great deal of work is still being done both in university departments and in government offices. And yet the confidence that was displayed through the practice of such research ten years ago has now been fundamentally shaken. No doubt a study will some day be undertaken which addresses the onset of this uncertainty as an historical phenomenon, but at a superficial level sufficient explanations are not hard to find. Amongst these must be included the shock effect of the publication of David Harvey's second book, *Social Justice and the City*, in 1973. A confident planning profession, continuing the development of its own academic roots as well as continuing the exploitation of its links with human geography and regional science, was suddenly faced with the prospect of an enforced return to first principles. As a result of this and no doubt many other factors, the last decade has been characterized by a fragmentation of research effort. Much analysis and investigation has continued – though probably expressed with rather less self-assurance – in the research traditions of the 1960s and early 1970s. Some research effort has been invested in each of a variety of radically different theoretical directions – from pure idealism to structural materialism. However, an increasing amount of effort has been expended purely at the level of epistemological debate. Moreover, this effort has steadily become detached from the realm of research practice. The early critiques, for example those of Sayer (1976) and of Harvey (1973) himself, were built upon the treatment of specific methods in planning or geographical analysis. They were almost autobiographical in the sense that the authors had clearly moved from traditional modes of research practice to more general levels of critical comment. However, within a very few years the more general debate had taken on a life of its own. Thus as early as 1978 Gregory was able to publish a contribution to that debate in which the vast majority of the references were to what may be loosely termed the 'epistemological literature' (whether within or outside the disciplines of geography or planning).

One result of this divergence of purposes was that most of the contributions to the debate written in the late 1970s spent little or no time spelling out the implications of their critiques at the level of research practice. What was in some cases stated explicitly and in others simply assumed was roughly as follows: since almost all traditional techniques and approaches are based upon a positivist model, and since that model can be shown to be fundamentally inappropriate to the social sciences, there can be no point in exploring the detail of those approaches any further. Quite apart from the dubious logic of such a conclusion, neither of the premises appears to be so self-evidently well founded as to warrant this wholesale rejection of traditional research and analysis practices. Part of the purpose of this book will be to help to bring the debate back down to earth, at least to the extent of attempting to explore a little more precisely what it all means at the level of the specific research questions which are repeatedly posed in a realistic academic or professional context.

A further result of the divergence of purposes noted above has been that contributors to the higher-level debate have, until recently, rarely attempted to bridge the gaps between the rival epistemologies proposed. It is not just that positivism is generally rejected (though indeed it is). It is rather that the more popular alternatives have each been cultivated in their own intellectual hot-houses and assiduously protected from any sort of corruption or compromise that might ensue through contact with others. This is perhaps legitimate as an exercise in the philosophy of science, but as a contribution to the advancement of our ability to inform the practice of applied research it is unhelpful. There is no indication at present that the intellectual turmoil of geography and planning will be speedily resolved in favour of one particular theory of knowledge, and so the most we may hope for is that something constructive may be synthesized from the main competing streams within the debate. Work has begun to move in this direction (see, for instance, Pred 1981, Gregory 1981 and Thrift 1983) but the practicality of the synthesis offered has yet to be demonstrated. The development of a rather different approach to the problem of synthesis, and one which is practical in that it builds upon the conduct of applied research, will thus constitute a second purpose of the following discussion. A third purpose of the book will be to compensate in some way for what is perhaps the most serious of the consequences flowing from the detachment of the intellectual debate. For it has not only bred a sort of fragmentation and polarization of standpoints, each cut off from a shared research tradition, but it has also failed conspicuously to stimulate major new research which is not just theoretically refreshed but is also located firmly and practically within the planning domain. Perhaps it is still too soon to expect each new perspective upon the acquisition of planning and geographical knowledge to have consolidated its own theoretical position in a manner appropriate to the support of applied research practice. Moreover, it is certainly not true to say that the high-level has had no impact upon the conduct of empirical investigation. Geographical and regional science studies are now appearing in both the materialist and phenomenological traditions. What remains unclear is how the majority of such studies are to be integrated back into the mainstream of policy formation and planning practice. Perhaps this would be a misguided aspiration, but it is clearly an issue which must be considered seriously. If the institutions and political imperatives of planning and urban policy-making constitute a fundamentally inappropriate context for the emerging research styles of human geography and planning, then the question of what is an appropriate context must be confronted as a matter of some urgency.

The problem is, of course, not just of academic significance. Planning as professional and political practice has come to depend more and more heavily upon a research base developed within the academic context of the human spatial sciences. It has adjusted its aims and its strategies to an approach to investigation which has been largely borrowed from the geographical and regional science methodological traditions. If those traditions are in the process of upheaval then the implications for planning practice are bound to be considerable. It is, however, much too simple to treat the problem as unidirectional. Academic research is both legitimized and refreshed through its contact with the practical and political spheres of planning and urban policy formation. Indeed, it is arguable that the epistemological debate which is beginning to have such a profound influence upon both the style and the content of research in human geography owes a great deal to an emerging political awareness amongst academic geographers, which may itself be traced to the increasing interdependence between their academic domain and its political counterpart in urban and regional planning. If this is indeed the case then it becomes all the more important for geographers to rethink their approach to research as a practical as well as a theoretical problem. The third purpose of this book will be to help to redefine this reciprocal relationship.

In the light of these purposes it may seem incongruous to adopt any method of argument other than by example. To explore the implications of an increasingly abstruse debate at the level of research practice; to look for workable analysis guidelines that bridge ever-widening epistemological gaps; and to attempt the reintegration of research – informed by such guidelines – into the political and practical frameworks of planning: all appear to demand exemplification through practice. There are several reasons why this book will not adopt that more obvious strategy. First, an empirical study, designed both to exemplify an approach to applied analysis and to satisfy its own internal requirements would be an enormous exercise. On the one hand, analysis techniques in planning and human geography span such a broad range of spatial scales, levels of aggregation and application areas – each presenting unique problems and requiring unique solutions – that no specific project could hope to encompass more than a small fraction of this range. Research at the interface between social and policy sciences is undeniably difficult to do. Research problems must therefore be very carefully specified if they are to be rendered tractable, and this requires the adoption of a highly selective approach.

On the other hand, an approach to applied analysis which is sensitive to the current epistemological debate is not likely to take on a simple and universally applicable form. As noted above, the debate at present seems closer to a complex stalemate than to a realistically workable resolution, and in such circumstances any attempt at synthesis must remain consistent with a variety of particular research strategies and techniques. It will not therefore be an approach which will be readily exemplified in a single empirical study.

However, even if such a straightforward approach were possible, it might not be the most appropriate one to adopt at this juncture. The debate has almost literally 'taken-off'. It has gathered its own momentum and is rapidly assembling a large family of active participants, each apparently quite willing to discuss the issue of planning or geographical analysis in essentially abstract terms. If, therefore, the debate is to be turned to useful practical purposes it must be addressed initially in its own terms. This does not mean adding yet another layer of pure theorizing, but it does mean treating the practices of applied research from a theoretical rather than a pragmatic perspective. In other words, we must start out from the main issues raised within the context of the higher-level debate and then draw them back down to the level of everyday investigative practice. The debate is not an irrelevant one but its specific relevance is in danger of being missed if for no other reason than through a breakdown in communications. The jargon of epistemology shares little in common with that of applied analysis.) If, therefore, this work is to fulfil the role of a 'cultural broker', it must make compromises in the direction of each culture. That almost certainly means starting at the level of philosophical debate, for it is that debate whose constructive and practical potential is unclear and untapped. The key questions – as to how we decide what research to do, and how it is to be done, when faced with the immanent complexity of urban society and its political imperatives – can only be answered by first clarifying the role of research in a political context.

The book therefore starts by attacking the problem both from the perspective of political practice and at a relatively high level of generality. In other words, it starts at the point of departure of those critiques of traditional research strategies whose interest springs from the context offered by planning itself (see for instance Camhis 1979 and Scott and Roweis 1977). Such critiques often begin by attacking the rationalist position, for this effectively neutralizes analysis and research, and thus renders analytical positivism logically feasible by imposing a rigid separation between political and investigative involvement in urban processes. Without exception recent critiques find this simple strategy quite unacceptable. The majority of these, however, attack the separation from a materialist perspective. Here we adopt a position which emphasizes the humanity of the subject matter and thus at the same time avoids the inevitably preemptive effect of many materialist critiques. In other words, the next chapter indicates the broad lines of an approach to applied research in planning and human geography which is lacking neither in sympathy nor in political awareness, and which remains consistent with a systematic reconsideration of the main techniques and theories of traditional approaches.

In the main review section of the book we pursue this more specific treatment by adopting a conventional and thus convenient means of classifying the basic techniques without claiming that it has more than a simple heuristic value. At the outset, therefore, techniques and approaches are distinguished purely on the basis of the level of aggregation of their objects. The discussion commences, in chapter 3, by considering some of the methods most commonly used to explore at a highly aggregated scale the economic and demographic bases of towns, cities and regions. Moving up the scale of spatial specificity, in chapter 4 we turn our attention to a range of simple techniques for the analysis of built form, or its aggregate pattern of use, when the issue of intra-urban or intra-regional locational relationships is assumed to be critical – as is the case with the demand for employment, for housing and for retail and related tertiary sector facilities.

All of the approaches discussed to this point share in common a primary concern with the behaviour of population aggregates. The theoretical object, even of a 'disaggregated' gravity model, remains the observable pattern of behaviour that emerges as the unintended result of the lifestyles of a large number of individuals. In chapter 5 attention shifts to techniques whose theoretical object is, no matter what analytical devices are subsequently employed, the observable or unobservable behaviour of a single individual or a small group (such as a family or the directors of a firm) which acts consciously and explicitly as a single unit. The review of these more micro-level studies commences with a discussion of techniques which focus directly upon individuals – upon the ways in which they understand the environment, upon the ways in which they impose meaning and value on it, and upon the ways in which they use as well as experience it in a selectively and reflexively non-random fashion. Finally, in chapter 6, we conclude the series of specific reviews by focusing upon techniques and approaches which address something of the full complexity of the issue of long-term locational choice itself. Whether made by individuals and households about their places of residence and of work, or made by the agents of firms and other institutions about the siting of production, distribution and service sector activity, such choices remain amongst the most central objects of applied research in planning and human geography.

In attempting to apply some of the lessons of the epistemological and theoretical debates at the level of research practice, the central review section will also be attempting to render those lessons accessible to those who find the relevant literatures impenetrable. It is unfortunate that the two literatures which are most in need of reconciliation (that of the epistemological debate and that of analytical practice in human geography and planning) are each characterized by styles and languages which are amongst the most difficult to follow. The presentation of the central section will therefore adopt, wherever possible and as far as is consistent with the overall development of the argument, a common simplifying format for each chapter. The treatment of similar sets of techniques and approaches to investigation will thus involve:

(i)   a brief methodological description, paying particular attention to underlying assumptions and principles, and where possible quoting examples both from academic and from professional research practice;

(ii) an outline of criticisms and limitations, differentiating those advanced within the frame of reference of the techniques from those which are effectively far more general (perhaps advanced from the perspective of the epistemological debate itself) and are thus as likely to question the frame of reference as they are to attack the detail of the techniques;

(iii) a synthesis section which draws out the implications of the two levels of critique and indicates the circumstances in which the techniques may be used legitimately in practice, and the ways in which their pertinence may be enhanced.

These synthesis sections will be consolidated at the outset of the concluding section, which will thus reconsider the limits of applicability of the various techniques and approaches discussed. At that point, however, we will drop the macro-micro typology and focus instead upon the crucial practical question as to which of these techniques has the theoretical ability to handle the processes of change at regional, urban or local scales. In the penultimate chapter, therefore, we reconsider and reclassify the techniques and approaches reviewed in the main body of the work, with a view both to isolating the necessary conditions of a dynamic mode of analysis and to highlighting the consistencies and inconsistencies of approaches that appear to meet those conditions. By returning to the principles enunciated in chapter 2, we show that there is more than one way of attacking the problem of interpreting urban change and that, though the alternatives stem from radically different epistemological traditions, they can contribute complementary rather than contradictory elements to a programme of dynamic analysis.

In the final chapter we return to a discussion of the planning and policy-forming contexts within which the various techniques and approaches discussed in the preceding chapters are required to operate. We return also to the theoretical issue which constitutes a main theme of the book – namely that of the detachment of epistemological debate in planning and human geography from a paramount concern with the practice of applied research. Drawing upon the conclusions of earlier chapters, the argument is finally proposed that a close attention to the practice of applied research is the only way out of a sterile polarization of the debate at an epistemological level.

Moreover, this reattachment renders the debate relevant both by defining the theoretical limits of current practice and by indicating the way towards a new practice which may rise, as it were synthetically, out of the ashes of the old. Furthermore, it is finally argued that if the intellectual roots of analysis in planning and human geography are appropriately reformulated, then the practice of policy formation may also be reconstituted, on a firmer theoretical basis.

Before developing this argument in detail, its central limitations must be acknowledged. It is intended as a general argument, applicable across a broad range of analysis techniques, and is advanced by treating specific theories, techniques and research studies as examples of approaches whose shared assumptions may be examined directly. The breadth of scope of analysis in human geography and planning is so great that the adoption of these non-specific terms of reference has meant, first, that some areas of applied or applicable research have had to be excluded from detailed consideration, and second, that the exemplifying techniques and theories which have been discussed have in most cases been the simpler and more popular versions. Thus, though some of the techniques employed in transportation analysis are discussed in what follows, the subset of transport research issues and problems is not treated as a self-contained area of investigation. Moreover, those techniques which are treated at some length tend to be the simpler forms. No pretence is made that a book as broad ranging as the one that follows could possibly do full justice to the most recent and most complex (and sometimes least accessible) research developments in each of the multitude of areas which constitute the field of analysis in the human spatial sciences.

The concern of the book is, in a sense, with the interface between the world of academic research and that of planning practice. The tensions that exist at this interface may be reduced by specific advances in particular research areas, but there is no necessary reason why they should be. The point is that those tensions can be examined directly, and for this purpose it is unnecessary to attempt the impossible task of scouring all of the many research frontiers which are, almost by definition, being advanced at points other than at that interface itself.

2

The scientific method in planning and human geography

The activity of planning is one in which we participate because we believe that today's decisions will materially affect tomorrow's welfare. Accordingly we attempt to bring a view of the future to bear upon those decisions, and to influence not only how things happen at present but how they will happen if our expectations and powers prove to be consistent. A planning decision, in other words, is one in which choices are consciously subjected to projective interpretation. It is such interpretation which will form one of the main subjects of subsequent chapters in this book. However, the bulk of this first context-setting chapter will be dedicated to a discussion of planning as a total process, of which investigation and interpretation form only a small part. Stated in such a blunt and brief fashion, however, the task appears almost limitless, since planning is a generic activity, applicable to a family trip to the coast just as to issues of urban policy formation. No attempt has been made to isolate well-defined decision or problem areas in which some particular form of planning may be appropriate. It might be reasonable for some purposes to start with the broadest of definitions and examine planning as a set of procedures independent of their context, but such an approach depends upon the internal coherence of the activity in question.

The point is that planning is such a fundamental activity that it has inevitably become infused with aspects of the subject areas in which it has been applied. It is thus impossible to pin down without either specifying its context or accepting discussion at a level of generality which would probably be of no more than semantic interest. This chapter opts for the former course. Specifically, it treats planning as a part of the process whereby public sector agencies intervene in the affairs of human society, whether through the building of bridges or via changes in the tax system. Moreover, its concern is with problems which manifest themselves at the subnational and local levels, rather than those whose symptoms appear national or international. The chapter will, therefore, attempt to illuminate these constraints by describing a set of procedures which, taken collectively, amount to the practice of planning within these prescribed limits, the purpose being to set up a discussion of the ways in which projective interpretation may be most appropriately pursued within such an overall framework. An approach to applied science may be derived which is thus both sensitive to its operational and political context and – with luck – effective at its own interpretative, explanatory and informative levels.

The limitations outlined above will naturally mean that the discussion which follows will not have general applicability. It will probably not be of particular relevance, for instance, to the problems of planning the production patterns of large-scale private sector enterprises. It will almost certainly be of little interest to those whose model of planning is a strictly professional one. Yet each of these areas is characterized by extensive coverage in the literature on planning and its information base. Ackoff's short treatise on corporate planning (1974) and Eversley's work on the role of the British professional planner (1973) are similar in that they each represent significant advances within their respective traditions. They are each critical of earlier studies and practices within those traditions. They each offer directions for change. Yet their critiques and propositions are in both cases heavily rooted in the traditions from which they spring. This is of course in no sense a weakness of the works in question, since any general commentary must be well rooted in the philosophical or practical traditions whose boundaries it attempts to extend. Otherwise it runs the risk of esoteric irrelevance. The point is simply that this set of discussions springs from a philosophical tradition – whose particular character will become clear as the argument develops – quite different from that of either of the texts referred to above. It will therefore have little to say to those planners whose roles and purposes are set by problems or styles of solution which exist in some quite separate domains. It is, after all, basic to the argument of each part of this book that the methods planners devise for both the interpretation and solution of problems must be sensitively tailored to the particular characteristics of those problems.

The chapter will, therefore, set a fairly precise and essentially applied context for the more specific discussions of techniques of analysis which are to follow. The obvious question, of course, is whether or not it is appropriate to discuss approaches to investigation in such a context. The concept of planning introduced above is that of a practical as well as political activity whose tasks include the manipulation of urban forms and institutions, and whose presuppositions include both practical knowledge and political imperatives. The book, in other words, starts out by treating the knowledge and information base generated by research effort not only as the product of a framework of certain academic theories and (possibly) scientific rules but also as the creature of distinct practical and even political interests.

As a discussion of research techniques in planning this approach will probably be found acceptable by most. However, to 'devalue' the techniques of human geography and regional science in a similar fashion, granting them no academically autonomous status, appears to require justification. After all, each represents a branch of one or more respected academic traditions – traditions that predate planning as a statutory ctivity by many generations.

It is, of course, the point of neither this chapter nor the book to denigrate the academic coherence or importance of geographical research techniques. One of the main points of both, however, is to adopt an explicitly applied stance in the treatment of research approaches in human geography and regional science, no matter what their academic pedigrees. The epistemological confusion which bedevils both of these disciplines at present is largely explicable in terms of a belated awakening to the unavoidable interpenetration of investigative and political activity. The ivory tower is no longer a sufficient base for any social scientific research, and the proposition holds as much for academically rooted spatial analysis as it does for politically triggered planning analysis. In this chapter we will explore the implications of this proposition (as well as reexamine its validity), but it is clear that it constitutes a starting point which is acceptable enough to justify approaching research techniques in the human spatial sciences – no matter what their origins   from the standpoint of their relationship with the broader political processes of planning and urban and regional policy formation.

The practices of planning

Any discussion of an activity as complex as planning must proceed by first adopting a procedure for simplifying the object of study. To interpret and explain the activity of football it would be necessary to discuss separately each of the rules, the relationship of opposition between the participating teams, the purpose, and each of the various tactics which individually or in combination may promote that purpose. Most of those who have received such simplified explanations and watched a few games should have a fair idea of what football is all about.

The same can hardly be said of planning, and this despite a veritable flood of learned texts upon the subject. Of course it is somewhat more complex and ambiguous than football, and so it is hardly surprising that our understanding of it is unsure. There must be hundreds of equally plausible explanations for our inability to capture the true nature of the process and describe it – many of which have been discussed in the literature. Yet one which is rarely aired, though it is at least superficially appealing, concerns not the ideology of the analyst nor the essential inscrutability of the object, but the manner in which simplification is attempted. We often assume that the method of the systems analyst is the only sensible way of simplifying a problem prior to searching for a solution. If we wish to find out how something works we take it to bits and examine each of its components separately.

This method has proved effective for biological and man-made machines, and it has been applied on many occasions to the analysis and explanation of social institutions such as planning. Its inadequacy in the case of patently human institutions like marriage or Buddhism has long been accepted, but its persistence as the recommended approach for the analysis of apparently more scrutable social units bears witness to its pedigree. Indeed, by far the most infamous caricature of the planning process – the classical theory of rational decisions – settles neatly into this category of what might be termed 'decomposition theories'. It attempts to illuminate and ultimately improve the complex activity of planning by first breaking it down into a related series of more basic social activities – establishing goals, doing surveys, analysing data, defining alternatives and applying criteria for the purpose of selecting amongst the options so defined.

The inadequacies of the particular decomposition quoted above have been extensively discussed. Probably far more important, and certainly far less frequently reviewed, are the problems associated with any approach to planning which attempts to achieve insights essentially by chopping it apart. Most of the modern attempts at a more realistic model of the process of planning drop the simple linearity assumption of classical theory, but still end up attempting to illuminate by fragmenting it into stages, activities, groups and techniques. The difficulties with such a strategy all boil down to one central problem. This is simply that if what makes planning a distinctive activity, coherently different from other complex social activities, is thought to reside in the patterns of relationships which it exhibits, then any decomposition theory is bound to fail. If planning is a coherent totality, then no matter how heavily one's composite flow diagram is embroidered with arrows it will barely scratch the surface of the institution in question. If the whole is indeed greater than the sum of the parts, we should be well advised to drop our penchant for pulling it to pieces in favour of a single-minded search for those qualities, attributes and symbols which lend it a distinctive coherence.

It may seem paradoxical to call for a holistic interpretation of planning in a text which, by addressing the topic of applied analysis and investigation, appears to presume the validity of an irreconcilably opposed approach. In fact there is no inconsistency since a central premise of this approach is that the planner's investigating and intervening functions are no more separable in practice than they are in theory. Moreover it is contended that this irresolvable fusion is one of the most important sources of the activity's coherence and distinctiveness. This claim cannot be effectively substantiated until the idea of an applied human science has been more fully developed. However, it is possible – indeed essential – at this stage to look briefly at the causes and consequences of attempts at rigorous separation and to build constructively upon an understanding of the situation in which we now find ourselves. For this purpose, and to initiate the discussion, we will first of all treat the practices of planning by adopting the traditional stereotyped distinction between investigation and intervention. We will then be in a position to attempt to build constructively upon its demolition.

Investigation

The urban planner has traditionally been accorded a clearly definable set of investigative functions which are treated as quite independent of his or her other activities. Such a demarcation is not difficult to explain. Whatever the historical sources of modern urban planning, the influence of the social sciences is unmistakable, and it is the methodological assumptions of these academic disciplines which have influenced heavily the way planners conceive of their role. Today probably the majority of urban planners are social scientists who have undergone some sort of conversion course to prepare them for their positions of responsibility. The academic and professional literature they read is for the most part written by social scientists (with or without the conversion course). The techniques they employ were for the most part devised by social scientists.

It would therefore be most surprising if planners did not assume that they had an independent investigative activity to perform, since that is what social science has been telling them, at least until comparatively recently, ever since the 1950s. The assumption has become about as basic as the belief that the word 'science' is appropriate when applied to the study of research topics within any one of the individual disciplines. It is thus as basic as the belief that members of such disciplines have a right to characterize their own research activities as purely investigative.

The intelligibility of the assumptions does not of course guarantee their validity. An enormous literature has accumulated since the debate was initiated by Auguste Comte, and it would be impossible to do it justice here. Nevertheless it is important, in a book devoted in part to an examination of the research base of planning, to be clear about the adequacy of the methods of the natural sciences which have been adopted so enthusiastically by planners brought up on a diet of positivist social science. If for no other reason, it is important because the reader has a right to know where the author stands so that the arguments of subsequent chapters may be correctly interpreted. At a more pragmatic level, however, it is also relevant since the methods incorporate built-in verification procedures. Thus if the assumptions necessary in transferring techniques from natural to social sciences are not warranted, these procedures may prove to be totally unreliable. As well as dictating verification procedures, natural science methodology, even when generously adapted for social scientific purposes, seriously limits the range of what may be employed as acceptable investigation techniques. It therefore also runs the risk of excluding from consideration potentially fruitful research strategies.

The scientific method as it is employed in the social sciences derives from two distinct and in some ways conflicting traditions (see Harvey 1969). On the one hand there is the inductive model, first described by Francis Bacon. This is built upon the logical concept of inference, since it argues that general rules may be inferred from a series of particular observations. The process of observation is itself supposed to suggest appropriate classifications. These in their turn enable generalizations to be made and so theoretical statements become the end products of a successful process of inference. Explanations and predictions are, in other words, said to evolve from experience via probabilistic inference. The problem with this approach is that it does not explain how purposeful observation can in principle be independent of a logically prior set of mental or intellectual categories: how in other words we can proceed deliberately towards general statements without first of all making fairly general assumptions with and through which to direct our study. This epistemological difficulty has concerned empirical scientists less than the practical problem of verification. The inductive method, in that it moves from the particular to the general, cannot avoid the necessity of a heroic leap which may be well informed but is ultimately based upon an article of faith. Just because I observe rainwater to flow down vertical drainpipes on many occasions does not, in and of itself, entitle me to start promulgating universal laws. As soon as I start generalizing I am in the realm of dogma.

The deductive model, due to Karl Popper, may be interpreted as a response to this difficulty – an attempt to overcome this problem of induction. It inverts the whole process by employing the logical concept of deduction. It argues that theory is logically prior to observation, that particular statements in an observation language may be logically derived or deduced from general theoretical statements or rules which are couched in terms of abstract concepts and non-specific categories. The generation of an initial theory is essentially an intellectual process. The deduction of specific hypotheses capable of empirical test through observation is a purely logical extension of the initial theory. A revised theory is the product then of this process of theorizing, deduction and empirical test. The problems with this approach are, as we should expect, the converse of those associated with the inductive method. Just as it is difficult to conceive of purposeful observation occurring in a theoretical vacuum, so it is hard to imagine where theory is going to come from if not from experience. Moreover, the deductive method only enables the analyst to falsify an initial theory. Thus a process of induction, or probabilistic inference, is required before any general or theoretical statement is accepted as true, since no theory can be subjected to every possible empirical test of its veracity.

In practice, all so-called scientific research adopts a strategy which mixes inductive generalization with deductive reasoning. Moreover, this approach seems to work fairly well, at least during periods of academic stability, within the natural sciences (see Kuhn 1970). Whether or not it is at all appropriate as a model for social science research is another question altogether. The issues are complex and have been reviewed by many social theorists including Giddens (1979), Filmer *et al.* (1972) Habermas (1972) and Berger and Luckmann (1967), each from a different viewpoint, but all critical of a simple positivist approach. It will therefore not be appropriate here to enter into a lengthy discussion of all the fine points of epistemology and ontology involved. However, it is important, for the purposes of this discussion, to refer briefly to the three most significant practical difficulties faced by the student of applied social science in attempting to adapt the methods of the natural sciences to his or her special needs. These may be conveniently characterized as the problems of interference, reduction and relativity.

The problem of interference results from the fact that the subject matter of research in planning and human geography – unique social events – is far too complicated to be handled through the application of the crude techniques appropriate to the study of the natural world. Social events and structures are ultimately the products of the deliberate and motivated choices of human beings who are capable of stepping back from, and thus confounding, any mechanistic generalization or prediction about their behaviour. Social or planning science thus become an inextricable part of the subject of study in a way in which natural science and the natural world never become so confused. Thus the traditional methods and verification procedures of the scientific method lose their efficacy when applied to more complex human phenomena. Not only do they fail to perform as they are supposed to, they also miss the point. Motivated social action requires the application of sympathetic techniques which capitalize upon the involvement of the researcher with the subjects of study. In other words, the social researcher only stands a chance of success if he or she deliberately stands the scientific method on its head by striving for subjective involvement with, rather than objective distance from, the subjects in question.

The second and related problem is that of reduction. Just as a systems theory approach to the process of planning is dangerous in that it attempts elucidation simply through decomposition, so a natural science approach to the phenomena of urban affairs can be equally problematic and dangerous for very similar reasons. The natural scientist generally believes that the subject of study can be usefully characterized as a machine. Not only does this behave according to simple stimulus-response rules – so avoiding the problem of interference – but it is also capable of investigation through decomposition since it is no greater than the sum of its parts. In social sciences, the analogous approach is that of methodological individualism which is based upon the assumption that human institutions are in principle reducible in that they are no more than the unintended consequences of individual actions. They thus exhibit none of the properties of an integrally coherent social unit, or a *Gestalt*. A debate upon this issue, initiated by Popper and Hayek in the 1940s, still continues (see O'Neill 1973 for a concise review). However, the most extreme position adopted by some individualists is now generally regarded as far too limiting a basis for social investigation. There is a real sense in which social institutions acquire a symbolic significance and thus take on a life of their own, influencing the ways in which choices are made and social life proceeds. Acceptance of this principle need not lead to the extremes of historicism and totalitarianism once so feared by individualists, since it is consistent, as we shall attempt to demonstrate, with a genuinely humanist applied science. However, such an acceptance does mean, as does an acceptance of the possibility of interference, that the mechanistic techniques of natural science will never be adequate to the needs of research, either in mainstream planning or in other areas of the human spatial sciences.

The third major problem, that of relativity, is more and more widely believed to be one that the natural scientist must face along with those working in social sciences or planning research (see Kuhn 1970). It is argued that any investigative activity is set within and dependent upon a framework of assumptions and values which collectively 'fix' not only the topics which are investigated but also the methods which are employed. Thus the results of any particular piece of work, however interesting and illuminating of the assumptive base from which they emanate, can never, except coincidentally, turn out to be fundamentally inconsistent with their framework. Kuhn believes the social sciences to be 'pre-scientific' in this respect. In other words, their practitioners face such great uncertainties that it is pointless to search for a single and generally accepted assumptive base within which all are happily working. This, of course, does not lessen the relativity of any particular set of findings. It simply reduces the probability of a full exploration of the potential of any one position.

In planning the problem is particularly acute because virtually all research is conceived and implemented within an avowedly teleological framework. It is not just an understanding we are after. We strive for a 'practical' understanding, one that will not only shed light upon a social problem but will also be suggestive of policy and design solutions. More often than not the subject of study is simply the relationship between a particular conception of a problem and a pre-selected set of potential solutions. Thus in a very crude and obvious way the political as well as academic assumptions and values which underpin the overall planning process inevitably infuse the research, so ensuring that those assumptions and values are continually reinforced.

Does this mean that the methods of the natural scientist are worse than useless if adopted for planning research? Certainly this is the most common conclusion in the recent literature (see for instance Harvey 1973, Buttimer 1974, 76, and Sayer 1976, 79). Moreover, if an espousal of the scientific method means a whole-hearted acceptance both of the Humean model of causal mechanism and the Popperian doctrine of methodological individualism, then scepticism is surely justified. However, it is by no means self-evidently clear that such extreme positions are the only ones which are tenable. Though the low-level operational guides are necessary adjuncts of the high-level epistemological paraphernalia of the hypothetico-deductive method, the converse is not necessarily true. Thus it is possible to adopt at least some investigational guides from the traditional scientific method without accepting its concepts of causality. This is the first constructive principle which may be drawn from the above discussion. Two examples will serve to illustrate the point.

First the Popperian method accords a fundamental structuring significance to a system of deductive logic. The importance is seen to be such that the method attempts further to exclude other, equally useful, modes of argument, and in so doing restricts its own utility. However, in acknowledging the inadequacy of the scientific method in this respect, it would be foolish also to deny the importance of a deductive system of logic. The internal coherence of an argument may be only a necessary condition of certain parts of social scientific investigations, but the intellectual discipline it has lent to studies in the past has been of enormous value, and should not be lightly discarded. The same is true of the falsification principle. An implication of the hypothetico-deductive approach is that theory is always either contingent or false. It can never be verified as certainly true. This has led Popper (1976) and others to build positively upon this logical paradox and suggest that the goal of scientific endeavour should be the falsification of theory: that scientists should always be attempting to derive empirical tests which will expose the inadequacy of their own intellectual edifices, since this is the only way that intellectual advancement will be achieved. For many much-discussed reasons, this unequivocal reliance upon the falsification principle is now seen to be misguided. But, as with deductive logic, the self-critical discipline it may lend to our research efforts should not be trivialized. Continual and diligent scepticism is a better state of mind to bring to established theory than, say, evangelical faith.

This example suggests a second more generally constructive principle which may be derived from the above discussion. If the mechanical simplicity of the methods of the natural scientist is denied to the social scientist primarily because of the humanity of the subject matter, then by the same token the humanity of the social scientist is likely to be the means whereby he or she is liberated from participation in a mechanically blinkered research process. In other words, what appears as one insuperable problem – namely the confounding self-consciousness of the human species – may be translated into a way out of another – namely the narrow relativity of all research findings. The social researcher, as a self-monitoring agent, becomes capable of stepping back from and developing a critical evaluation of the assumptions, political values and methodological proclivities of his or her own investigational style. Just as the subjects of any piece of research can invalidate or fulfil any prediction about their behaviour simply by choosing to do so, so the social scientist is equally capable of escaping from any particular ideological prison whose limiting features can be described, discussed and deliberately evaluated. All that is required is a process of creatively self-aware speculation. Thus the dialectical materialist can become a structural functionalist and the structural functionalist an idealist.

The final constructive rule that may be extracted from a critical treatment of the natural scientist's approach is an avowal of the virtues of inductive sensitivity in research design. It may seem obvious to suggest that techniques should be tailored to the idiosyncrasies of particular tasks, but every attempt to devise a fully comprehensive model, a meta-theory or a ubiquitous method is a denial of this rule. One of the purposes of this book is to demonstrate the importance of such a sensitive approach. It should not, however, be presumed that this will inevitably mean the reduction of social and planning science to an arbitrary programme of opportunism, even though this seems to be the direction in which some research initiatives, discussed in the next section, are leading. Techniques may be sensitively devised and still be consistent with certain fairly basic principles. They may for instance be required to follow the low order 'investigational guides' of the traditional scientific method referred to above. More important, they most certainly should be required to respect the humanity of their subject matter. However, within such a framework, which will be elaborated further in subsequent chapters, it is of considerable importance that techniques should be applied, not in an unthinking fashion or through the application of generalized mechanical rules, but by sympathetically structuring each research strategy about the unique features of the problem in question.

Of course these are not the only principles that should inform investigations in urban planning and human geography. The argument to this point has done no more than lay a few ghosts. All that has been demonstrated is the fallacy of a view of applied social research as a 'neutral' and external vehicle for the provision of 'objective' information upon which 'unbiased' policy judgements may be built. This image has been shown to be false through a very brief and mostly theoretical look at the research activity itself. The major part of the remainder of this work will be devoted to a much more detailed and practical examination of the activity of applied social research, out of which we shall attempt to draw more substantial guiding principles. These principles will address several of the key epistemological issues left as yet unresolved in the above discussion – issues such as the gulf between humanist and materialist positions which still appears unbridgeable (notwithstanding the heroic attempts of the self-styled structurationists who have built upon the work of Giddens 1979, Bourdieu 1979, Bhaskar 1979 and even Berger and Luckmann 1967). The rejection of naïve positivism does not solve these problems, nor does it obviate the study of research practice.

However, before the context of such an examination is complete, it is necessary briefly to shift our perspective. Instead of moving at once to a discussion of the research practices themselves, it will be appropriate at this point to treat them just as parts of the wider political and administrative process in which they persist. This is particularly important in view of the second of the two problems noted above as characteristic of an attempt to adapt the scientific method to the needs of the planner or urban policy-maker – namely that of the parochial relativity of any findings produced. If this criticism is pertinent, and some would now accept it even when levelled at research in the natural sciences, then it immediately calls into question models of the planning process which depend upon a strict separation of functions – the decomposition theories alluded to earlier.

Intervention

The classical theory of rational decisions (succinctly described and elaborated in Dror 1968) is the crudest and simplest of the decomposition theories, though it has undoubtedly spawned many far more sophisticated derivatives (see for example McLoughlin 1969, Catanese and Steiss 1970, Chadwick 1971, etc.). Never intended as a description of any particular decision process, it is worth considering even as an ideal type if only because it demonstrates with great clarity both the aesthetic appeal and inherent danger of the attempt to separate the functions of intervention and investigation. Figure 2.1 below is drawn so as deliberately to highlight the separation.

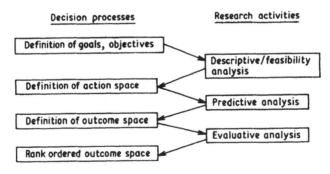

2.1   The rational decision process

Each of the various sorts of investigative exercise are seen as external to the decision sequence, at most facilitating the passage from one stage to the next. When information is required it is drawn in to provide unambiguous answers to the various substantive questions which are raised in the course of reaching a rational decision. Thus, once the goals and objectives of planning are precisely fixed through the independent political channels of representative democracy, decision areas will naturally emerge. Each can then be neatly circumscribed by applying a variety of descriptive techniques which completely fix the bounds within which choices may be made. Each feasible option is then extrapolated into the future of the real social world, this time by drawing upon predictive techniques – and an outcome space is defined to correspond with the action space. Objectives are then applied to each outcome, through the application of evaluation techniques, and so a rank order and an optimal decision is achieved.

The practice of planning and policy-making is, of course, not like this at all. Research into the process of planning – notably in the UK the work of John Friend at the Institute of Operations Research (see Friend and Jessop 1969 and Friend, Power and Yewlett 1974) – suggests that the rational model is so far from any sort of realistic description that its value even as an ideal type is called into question. Goals and objectives are never clearly articulated at the outset, but evolve as the process unfolds. The trigger for action is more often either legislative necessity or a snowballing perception, from a particular and unstated point of view, that a problem exists. The action space is never neatly bounded, for in a policy-forming or decision situation, the presupposition of only modest originality upon the parts of the actors is enough to ensure a theoretically infinite variety of potential solutions. In practice, the problem is usually reduced, arbitrarily and necessarily, to a set of two or three feasible solutions which may be evaluated predictively. The prediction stage, for the sorts of reasons mentioned in the preceding section, is fraught with difficulties, and systematic evaluation is impossible since objectives are neither clearly formulated nor practically convertible into a cardinal currency.

The alternative models offered fall typically into one of two distinct categories. One group tends to drop altogether any attempt at examining the processes of planning and policy formulation directly, concentrating instead upon their economic and political contexts and outcomes. The other invests much effort in investigating those processes, and suggesting incremental improvements, but generally paying attention only to their most immediate contexts and most proximate outcomes. Most of the recent Marxist critiques fall into the first group. Best known is probably the work of Castells (1972) and its subsequent elaboration in Scott and Roweis (1977). Both adopt an essentially Althusserian structuralist position. According to this view, no branches of the state, and this includes urban planning, have any real autonomy whatsoever. They are all intimately involved in the pattern of relationships that characterize the social system of the time. If this pattern is dominated by the relationships of the ownership and control over the means of production, as in classical Marxist theory, then the role of the state as subservient to the interests of the property-owning class becomes not much more than a reinforcing one. The activities of the state are seen as achieving little more in the last analysis than the reproduction of labour power. Their purpose within the capitalist scheme of things is simply to provide the houses, schools, hospitals and roads which ensure that there exists at all times an available, competent and healthy work force. A strict structural account such as this can find no use for detailed empirical studies of either the actors or the practices of the various branches of the state. The details are irrelevant since the outcomes are completely determined by the structured pattern of relationships that pertains.

It would be impossible, within the scope of a relatively specific account such as this, to do justice to the rapidly growing body of literature which addresses the argument briefly outlined above (see for example Cockburn 1977, Broadbent 1978, Harloe 1977, Tabb and Sawers 1978 and Dear and Scott 1981). One general point is, however, worth making. It is simply that what I have termed applied human science cannot possibly be built exclusively upon an analysis such as the original work of Castells. Althusserian structuralism shares with positivism a penchant for a deterministic model of causal mechanism. In other words, they both treat social behaviour as explicable in purely contextual terms. The physical, social and economic circumstances of the individual are all we need to know if we wish to predict his or her pattern of activity. Planners, like other government employees, are incapable of breaking out of the framework of class affiliations and role reinforcements which collectively determine their overall impact upon social affairs. This position is patently inconsistent with a view of the individual as a self-conscious and creative agent. Within this alternative intellectual framework, the planner, just like anybody else, is regarded as capable of deliberating creatively and reflectively upon any given situation, and making an effective choice as a result of such reflective deliberations. The rationale for and implications of this essentially humanist position will be discussed further in subsequent chapters. It is relevant here simply to re-emphasize its inconsistency with a strictly structural interpretation of anything, including the role and efficacy of the planner.

The second response to the patent inadequacy of the strict rational model is that of the process analysts. Rather than starting with the socio-economic context of a political decision process, they concentrate upon examining its internal complexities. They typically say very little more about the context than that modern urban policy-forming practices are a necessary response to the complexity of that context which is seen as a daunting and multi-layered patchwork of interdependent decision areas. The suggested response to this 'inter-corporate' complexity varies from author to author. However, the answer of the best-known group, referred to above (see Friend, Power and Yewlett 1974 and Friend 1980), is a process of 'strategic choice'. A strategic choice is said to be required whenever separate decision areas are highly interconnected. In such a situation a recursive strategy is essential, as is a generally flexible and adaptive approach to problem-solving. This leads to the breakdown of clearly defined tasks and the inevitable development of reticulist skills amongst so-called 'professional' planners. Research becomes a purely heuristic device, to be applied in any way that works when uncertainty cannot be managed in a quicker or cheaper fashion. Its independence from other planning tasks thus disappears altogether.

The apparent paradox of this response is that, though it emanates from an intellectual tradition which is quite different from that of the Althusserian structuralists and adopts a totally different view of the context in which decisions are taken, it can be rendered consistent with the approach of materialist urban sociology without much difficulty. In practice the process analysts use their characterization of the strategic choice framework as a base upon which new and 'rational' (in a non-classical sense) problem-solving and analysis techniques can be built. However, in theory their view of the way in which planners 'necessarily' respond to a complex and interconnected decision environment is one which lends itself ideally to incorporation within a wider instrumentalist treatment of the state. For the process analysts' model breaks down the fundamental separation of investigation and intervention, and so removes that link with an objective reality which lent the classical rational mode its apparent neutrality. The rational planner could remain above the fray purely because he or she was also a scientist in the traditional sense. Indeed, an illuminating way of viewing the rational model is as a means whereby objective truth can, by the simple application of an unambiguous criterion of optimality, be translated into action. Once the possibility of objective truth is denied, then no matter what the mechanism of the decision process, the criterion of optimality becomes critical.

No doubt most of the process analysts would reject a Marxist interpretation of the economic and historical context, and so would certainly not accept that the criterion for urban policy formation in a mixed economy amounted in the final analysis to little more than an injunction to maintain the *status quo*. However, in that they tend to ignore this economic and historical context, and at the same time describe a decision process which exhibits no obvious built-in checks and balances (such as rationalists believed to be offered by the scientific method), their approach is in one very clear sense consistent with that of the structuralists. The adaptive, recursive and reticulist process they describe lends itself ideally to all manner of conscious and unconscious manipulation – in the interests of international capital or whatever other dominant force or collective unconscious is believed to hold sway over the progress of human affairs.

It is not surprising in these circumstances that moral philosophers and political theorists have continued their search for fundamental rules of social organization; principles of such intellectual, political and ethical force that they can form a sufficient justification of and criterion for the intervention of the state in the lives of individuals. The importance of such a search for planners must be obvious. If the argument of the preceding section is accepted, then planning has lost its methodological *raison d'être*. It can no longer rely upon the neutrality of information to guarantee its independence from the purely political activities of the state. However, this need not be a problem. If a criterion of optimality can be found which is so pre-eminently just and reasonable that it is accepted by all members of a social unit, then it can act as the arbitrating focus for public sector decision-making in the increasingly complex inter-corporate sphere described by John Friend. The fact that the social complexity of that sphere (amongst other things) precludes the separation of political intervention from 'scientific' investigation is irrelevant so long as both functions are infused with the same concept of justice and principles of optimality. The net effect of researching, designing and implementing urban policies through a daunting variety of interlocking and overlapping public sector agencies will be both maintenance of the accepted level of individual liberty and continual advancement towards the desired distribution of social welfare.

The bitterness and persistence of the debates between supporters of alternative 'universals' in the domains of ethics and politics does not fill one with optimism that the search will soon be complete. Ever since Rousseau and Kant the literature on 'social contracts' and 'fundamental principles' has grown continuously. Each author adds his own distinctive perspective to the debate, and so it becomes increasingly difficult to summarize. However, three of the currently most popular positions will be reviewed briefly here in order to give an indication of the breadth of the spectrum.

The most conservative principles defining legitimate spheres and criteria for state intervention, such as those of Kant, have long since been overtaken by events. All governments have now extended their areas of intervention way beyond what such theorists would accept. However, the individualistic principles underlying the various versions of the doctrine of utilitarianism do offer an essentially conservative criterion of optimality which is still very strongly supported. The policy-guidance precept which derives from this doctrine is, in its simplest form, that that policy should be selected which will maximize aggregate benefit. It is a conservative principle in that it allows a greater utility or benefit to outweigh a smaller one. And it is individualistic in the sense that its justification is based upon extending an intuitively appealing principle of individual action to the realm of the state. Individuals, so the argument goes, generally and in the long term attempt the best for themselves and their families. They defer gratification so as to let greater benefits outweigh smaller ones. Why therefore should not states behave in a similar fashion? This argument, advanced by Bentham (1789), Mill (1867) and many political theorists since, was subsequently adopted by neo-classical economists as the Pareto criterion. A more widely applicable version is the Hicks Kaldor compensation principle, which introduces money as a way of getting around certain technical difficulties (see Mishan 1969 for a review of the debate in welfare economics). This extension of utilitarianism argues that policy A is to be preferred to B if the beneficiaries of A are willing to pay the beneficiaries of B enough to persuade them to agree to A. Though superficially appealing, even the Hicks–Kaldor version of the doctrine of utilitarianism is unlikely ever to achieve the status of a universally acceptable criterion of optimality. For a start, its systematic implementation, even if full compensation is paid by beneficiaries to losers, would mean an overall redistribution of welfare which was regressive. The rich would always be willing and able to outbid the poor since their marginal valuation of the compensation currency would be lower. Moreover, and more generally, the doctrine fails because it ignores all the collective properties of society. Even if the principle is apposite for the individual, societies are not analogous to individuals simply because they are composed of them. They would have no meaning for their members if that were in fact so.

An alternative principle of justice and optimality which springs from a liberal tradition but attempts to build constructively upon this last criticism is the 'maximin rule advanced by John Rawls (1972). The optimality criterion offered in this case is that policies should be selected which, consistent with certain basic liberties, maximize the benefit of the least advantaged group in society. All other groups are ignored. The rationale is quite different from those offered by the individualistic utilitarians. Rawls postulates a hypothetical social contract, a state in which each individual knows nothing about where he or she is likely to end up in the overall social welfare distribution. In such a situation it is argued that a rationally self-interested individual would play it safe and opt for the maximin policy on the assumption that there was a significant chance that he or she would be at the bottom of the pile, needing the benefit of redistributive state action. Apart from problems of definition (about precisely which is to count as the least advantaged group in society) the most frequently voiced reservations with Rawls' principles of distributive justice relate to the central rationale which cites a notional contract. Why, for instance, should rational people be assumed to be so unwilling to take risks? The ethically appealing features of the central principle tend to get lost in a rather unproductive debate about a most unreal situation. Quite apart, however, from the concept of a hypothetical contract, it is very important that we should not in fact lose sight of the implications of the principle itself for it does stop short, in focusing upon only one end of the social welfare distribution, of advocating long-term equalization. If for instance the poorest are assumed to progress fastest in a situation in which differentials are maintained, then the maximin rule would support the maintenance of those differentials.

The only criterion that focuses explicitly upon differentials rather than the absolute welfare position of the poorest, and thus upon eventual equalization as its political goal, is one which might be derived from certain materialist or pure socialist ideologies. In practice, at least until very recently, it has been hard to find explicit commitment in the literature to fully fledged Marxist criteria for urban policy formation in a mixed economy. This is no doubt partly because such a commitment would amount to an attempt to derive principles of justice and state activity for a given political context from an interpretation of human society which was inconsistent with the persistence of that context. Certainly a Marxist criterion could not seriously be put forward as the basis of a political consensus in a mixed economy since it emanates from a pure conflict model of society within which no meaning can attach to the concept of universally or generally acceptable criteria for public sector action. However, it is interesting, if only for comparative purposes, to speculate about what sort of pure socialist criterion might be offered. Expressed in terms analogous to those of the

maximin principle, it would presumably endorse only policies
which maximized the reduction in differentials between the
most and least advantaged, no matter what the absolute effect
on the least advantaged. It would focus, in other words, not as
in the Fabian literature upon equality of opportunity, but upon
equality of outcomes. Such a criterion follows clearly, for those
of an egalitarian disposition, from research such as that of
Jencks (1972) which suggests that even equal access to education
is insufficient to remove persistent discrepancies in income and
satisfaction levels. Thus it may continue to gain support even
amongst those to whom the analysis typically accords a role of
unequivocal opposition.

One very simple way of illustrating the dilemma posed by
such divergent optimality criteria is by constructing a simplified
hypothetical choice. If the society in question – whether a city, a
town or a nation – is characterized by only one significant
division, namely that between the rich and the poor, and the
welfare outcomes for each group of alternative policies can be
precisely quantified, then policies can be compared using a
simple histogram as in figure 2.2 below.

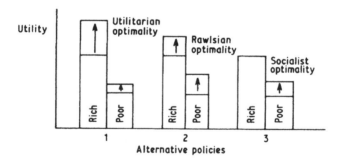

2.2   Policy evaluation given alternative optimality criteria

Assuming the three options under consideration yield
quantifiable benefits to the rich and the poor as indicated by the
arrowed portions of each vertical bar, then the choice depends
purely upon one's preferred optimality criterion, or as Rawls
terms it, principle of distributive justice. Policy one maximizes
aggregate benefit, but fortuitously only through the enormous
gains of the rich. Policy two maximizes the gain of the least

advantaged, but only through an increase in the gap between the two groups. And policy three achieves a reduction of differentials, but minimizes the overall expansion in welfare.

Liberal democrats would seek a way out of this apparent impasse through the mechanism of social choice rather than any particular decision criterion or justice principle. The way forward which they might espouse would be so to democratize the process of policy formation that there were checks and balances sufficient to compensate for the worst excesses of inter-corporate complexity, no matter what other principles might be consciously or unconsciously adopted. In other words, public participation would be extended, formalized and made decisive. The ever-increasing political activity of pressure groups – whether trade unions, amenity societies, or residents' associations – bears witness to the growing belief on the part of many people that they have a right to participate actively in a public sector decision whenever it appears that their interests may be materially affected by its results. A quinquennial voting choice between particular individuals who represent generalized and pre-packaged interests is no longer seen as sufficient.

The perceived sufficiency or otherwise of representative democracy is, however, not the issue. No matter how popular the idea of a participatory democracy becomes it will still remain relevant to ask of its procedures whether or not they fully legitimize all the activities of the state. It will still be relevant, in other words, to question the principle of substituting citizen participation for society-wide rules of distributive justice, especially as there appear to be at least two very strong grounds for doubt. The first of these was originally outlined by Arrow (1951) and has been much debated since (see for example Little 1952, Samuelson 1967 and Sen 1970). The argument has never been faulted, however, on technical grounds. It shows that for any given choice in which there exist more than two alternatives, there can be no simple voting procedure which will guarantee the logically consistent translation of individual rankings into a social welfare function without violating very appealing ethical principles. The logical principles upon which this conclusion rests are that the translation be 'collectively rational' and 'independent of irrelevant alternatives'. Collective rationality simply means that if all else remains constant save that one person ups his rating of one option, it should at least not fall in the social order. Independence of irrelevant alternatives is equally straightforward, in that what is required is a process of social choice which depends purely upon the rank orders adopted by each individual from amongst the alternatives available. It should not also depend upon some extraneous device such as a yardstick or cardinal scale. The ethical principles advanced are the Pareto rule, that if all individuals prefer A to B so must the social welfare function, and the non-dictatorship assumption that no single individual's preferences predominate in defining that function. The proof of the impossibility theorem need not concern us here, since all critics seem to accept its technical validity. It builds upon Condorcet's much discussed voting paradox which occurs in the situation in which three options are ordered by three voters respectively, ABC, BCA and CAB. In such a situation there is a majority available to vote A over B, B over C and C over A. An obvious way to resolve the paradox is to ask each of the voters to declare the strengths of their preferences. This at once introduces 'irrelevant alternatives' in that some sort of cardinal scaling device must be employed. Such a way of relaxing Arrow's rules has a particularly unfortunate pertinence for this discussion since it amounts to an acceptance of the utilitarian criterion and converts participatory democracy into a technique for finding the maximum aggregate benefit option. Thus if social choice is to be based upon a voting procedure in which individuals reveal the cardinal strength of each of their preferences as well as a rank order, then that choice can be consistent with the utilitarian criterion and with no other.

Participatory democracy is thus seen to be dependent, if it is always to be effective, upon a logically prior principle of justice or ethics. The immediate question raised is: why this particular principle? We could equally well relax Arrow's non-dictatorship rule and make our democratic procedure dependent upon the ethical principle that only the votes of the least advantaged shall be counted. In other words our mechanism of social choice is rendered consistent with Rawlsian rather than utilitarian optimality. Alternatively, we might wish to question the validity of any attempt to translate even a subset of individual values into a social choice. Let us assume for the moment that unambiguous choices could be derived from original voting procedures. A Marxist would still no doubt argue that there was nothing self-evidently acceptable about the 'dictatorship' of the majority brought about as a result. He or she would point to the pattern of social and economic relations which were more than sufficient to 'determine' the long-term distributional effects of any participatory democracy when overlaid upon a capitalist economic structure.

Whether the problem is seen as fundamental to the structure of capitalism or as purely the result of the inter-corporate complexity of modern urban policy formation no matter what the pattern of ownership, a problem must certainly be acknowledged. Local participation rates are generally found to be low and positively correlated with class and income (see Cole 1974 and Stringer and Ewens 1974). More important, the chances of materially influencing a public sector decision are also class related (see Wolpert *et al.* 1972). When participation exercises are required by statute – as is the case with structure planning in the UK – they operate at best as sensitizing devices and involve no devolution of decision-making power. Most in fact fail to achieve even this level of involvement and operate as low-budget and fairly ineffective public relations exercises (see Goldsmith and Saunders 1975). In such a situation planning and policy analysts are cut adrift from virtually all means of political absolution. It appears that not even recourse to the model of the Greek city-state is sufficient to ensure them a purely technical and neutral function. Even such a model can be shown to depend upon logically prior ethical principles if its unambiguous operation is to be rendered theoretically safe. Moreover the reality of participation practice has proved woefully disappointing, leaving vast areas of discretion and influence to local officials.

Conclusions

It will be helpful here to draw together the various strands of the argument as it has developed to this point, if only to demonstrate the impossibility of easy answers. The task is essentially one of devising an overall interpretation of the practices of urban planning and policy formation which accords a coherent role to the activities of investigation and research. Of course this is a purpose whose instrumental rationality with respect to work on research methods is obvious. From the point of view of the student of or practitioner in planning it may, however, seem to get the cart before the horse in that a clear role for research is demanded of the more basic political activity, rather than allowing that role to emerge or not as the case may be from a realistic characterization of the context and practice of planning. Both responses to the over-simple rational planning model discussed above in fact chose the latter approach. Althusserian structuralists interpret the state in terms of a model of society, planning in terms of this model of the state, and planning research purely as a further device operating to reinforce the whole process. The process analysts generally but tacitly adopt a different model of society, go into far greater detail in describing the practices of planning, but still accord research a very modestly heuristic role. The approaches are, however, similar in that neither attempts to save planning for the rationalist by hanging it on the peg of a neutral and objective research and information base.

Of course the implication of accepting the arguments of the first main section of this chapter is that this similarity between the two approaches is readily justifiable. Planning and policy analysis, just like any other research in the social sciences, cannot afford to ignore the special and human nature of its subject matter. Rather it must capitalize deliberately upon this uniqueness, devise sympathetic and involvement-generating techniques, and so sacrifice completely the neutralizing and objectifying benefits of the separation of subject and object. However, this should not mean belittling or diminishing the role and importance of research in planning. We have seen that there is no way of saving either the planner or the social researcher by recourse to that which is external to the tasks they must fulfil. Just as there is no such thing as a neutral and objective research method, so there is no such thing as a universally acceptable principle of distributive justice, criterion of optimality, or rule governing state activity. Moreover the extension of community control via new mechanisms of participatory democracy seems both in principle and in practice to offer no real alternative way out of the relativity trap. The answer is probably not even to attempt to escape the trap but to make a virtue of it: to build a model of planning and policy formation upon the principles of a human science rather than to downgrade both; and to work towards a mode of research practice which is consistent with the design of sensitive and humane urban policies.

Before proceeding to attempt a start in this direction, it is important to be clear about the implications of such a venture. In arguing that a coherent and pertinent model of public sector planning can be built upon a particular conception of social scientific investigation, we are not devaluing its political and regulatory aspects. We are rather suggesting first, to summarize the main point of this chapter, that they are indissoluble, and second, to preview the remainder of the book, that investigation must be tailored to the humanity of its subjects and plans to the products of such investigation. This does mean rejecting, at least as ideals if not as descriptions, all models of planning which treat it as purely instrumental to some dominant political philosophy or ideology, and all models which treat it merely as a control process in a state of continuous adaptation to the complexity of the objects of control. By the same token also it means rejecting all models of applied research which in turn characterize investigation as totally dependent upon some overweening purpose, whether this be a grandiose, political and ideological one, or just the mundanely pragmatic aim of making awkward and pressing problems go away.

More positively, it means building upon the critically constructive approach to social scientific research in planning and applied human geography outlined earlier in this chapter. It was argued there that such research could usefully accept three principles emerging from an essentially humanist critique of the practices of modern social research. First, some of the low-level operational guides of the hypothetico-deductive method – such as the demands for internal logical consistency in theoretical statements and for constant scepticism implicit in the falsification principle – appear well worth retaining. Moreover the similarity between subject and object, which renders dangerous many of the usual techniques and rules of scientific method, can be translated into a positive advantage. If the reflexive and self-monitoring capacities of human beings are viewed, not just as the sources of social science's inability to predict, but primarily as the means whereby the social scientist can cope with the limitations imposed by his or her own ideological position, then a way forward is clearly indicated through constant self-criticism. Finally, the rigid methodological dogmatism associated with an uncritical acceptance of the seductive consistency between particular purposes and techniques should be assiduously avoided. Rather techniques should always be sensitively tailored to subject matter. This is not a plea for a return to undiluted pragmatism since the subject matter will always and ultimately be human, and this single principle is the one which, for this book, is both beyond question and at the same time the source of all questions.

Important though these principles are, as yet they amount not even to a definition of a programme for applied social research, let alone to one for the total and complex political and technical activity of planning and urban policy-making itself. In order to progress towards the goal of humane urban policies, we must proceed cautiously and slowly. We must first emphasize that to build a view of the essentially political activity of planning upon one of the apparently narrower activity of social research need not be an immodest goal. It involves simply the acceptance of two principles: first, that the interpretative function of social research can never be purely technical but will be reflexively rooted in the human and social processes it observes, and second, that the political function of intervention is, amongst other things, a part of this reflexive process which is as logically dependent upon interpretation as interpretation is upon action.

In other words, it is not really a question of devising grandiose programmes of research and political action. The problem is to ensure that each planning decision that is made, each new policy that is formed, bears the stamp of an appropriate intellectual and ethical pedigree. No matter whether its purpose rationality is Marxist or utilitarian, in terms of this model it will be a better decision if its research base is both reflexively critical of this rationale and openly sensitive to what it will mean to those whose lives may be materially affected. The remaining chapters of this book will be dedicated to examining the various methods and domains of human geography and planning research that can potentially provide such a base.

What we have achieved in this chapter is the initial definition of a context broad enough to span the apparently independent worlds of planning practice and academic social research. For we have seen that neither is independent of a framework of values and both are dependent upon an approach which acknowledges the humanity of the object – whether that object be the target of political action or the topic of investigative scrutiny. The central point is that the subject (the planner or human geographer) and the object (the client, the citizen, the actor or the respondent) stand in intuitive or sympathetic – as well as physical – relationship one to the other, and in each case this is the critical activity defining relationship.

For the next section of this book, therefore, we shall effectively ignore the distinction between applied and academic social research, between planning analysis and human geography. Whatever the technique or approach to investigation under scrutiny, the method of presentation will in each case be similar. We shall start with the presupposition that, whatever else may be overtly or covertly intended, analysis in planning and human geography serves initially at least simply to reduce uncertainty. We have been concerned in this contextual chapter with more esoteric issues, issues to which we shall return throughout the more specific discussions which have to follow. Moreover, the notion of analysis as uncertainty reduction is not intended as a means of reintroducing any sort of objectively neutral peg upon which a pretentious pseudo-scientific analysis framework may be hung. The questions about which sorts of uncertainty should be reduced, how and for whom, do not disappear as a result of organizing our discussion in this way. Nevertheless, it does provide a modest lower-level framework within and from which higher-level questions can be more precisely distilled. Thus we shall examine certain broadly defined areas of uncertainty in which basic information may be appropriate to the needs of planners and the interests of human geographers. We shall in each case assess the relevance – from the point of view of an applied human science as outlined above – of research in these areas. We shall look briefly at the techniques available, the assumptions they make, how they are typically used, and how they might be improved and applied in ways which are more reflexively sensitive and less mechanically rigid. And we shall conclude by returning to the broader question of how such attempts at human planning and geographical analysis should be expected materially to affect the plans and policies to which they are inevitably related.

3

The economic and demographic bases

The routine core analysis of a community's aggregate economic and demographic base hardly seems to admit of testing against the sorts of criteria for applied science which we began to develop at the end of the last chapter. The design of such an analysis is typically so far removed from the level at which individuals actually operate that it seems strange to expect its techniques and procedures to be transparently sensitive to the attitudes and perceptions of those individuals. Moreover, the analysis strategy normally involves the use of such commonplace curve fitting and data management devices that it seems equally strange to demand that its practitioners should be constantly and critically aware of the purposes that those techniques presuppose. It is almost like expecting the production of a geological map to be discussed in detail with all who walk upon the hills – presumably by cartographers who are also constantly assessing the political significance of the colours they are using!

It is, of course, easy to caricature any attempt to set such apparently commonplace practices within a framework of intellectual and ethical principles. The discipline of actually adhering to those principles is much harder, and probably quite fruitless unless the practices are well understood as activities each with their own specific purposes. For this reason, it is relevant to examine precisely the manner in which even the most aggregate analysis techniques attempt to reduce specific domains of uncertainty before examining the broader implications of their use. The ethical and intellectual principles relate to the meaning which may be extracted from alternative ways of using techniques. Yet each of these techniques is idiosyncratic at least with respect to the aspect of an uncertain past, present and future to which it is addressed. Its meaning in use must also therefore be to some extent specific, and accessible only through an examination of that use. This chapter will thus attempt a constructively critical treatment of what are undoubtedly very mechanical approaches, still adopting the point of view of an applied human science, but applying that point of view, as it were retrospectively. We shall look at the two basic areas of demographic and economic uncertainty, at specific investigative and projective techniques which seem most likely to reduce uncertainty in each of those areas, and then examine the implications of using such techniques in terms of the intellectual and ethical principles outlined towards the end of the preceding chapter.

Before examining the analysis problems encountered in addressing the community's economic and demographic bases, however, it is well worth pausing briefly to consider the importance of the tasks. It may well be felt that, even at the regional level, since most of what determines the shape and dynamics of a community's economic and population structure is way beyond the influential reach of its planners and policy-makers, sophisticated analysis of these factors will not be worth while. Such a viewpoint seems, for instance, to be quite consistent with what I have termed the process analyst's model of urban planning. If each planning exercise is thought to be conditioned essentially by some initial problem perception then it will only be when, say, unemployment or a falling birth rate are regarded as locally soluble problems that basic economic and demographic analysis will be required. Since these are not generally so regarded, it would appear that such analysis will not generally be seen as cost effective.

Such a view ignores the basic complementarity between the investigative and interventionist functions of planning. No matter how analysis is affected by and in its turn affects the policy forming and implementing aspect of the whole process, so that they both become locked together in a mutually reinforcing framework, at one low level at least they remain different sides of the same coin. No matter what overall purpose they both reflect and refine, that purpose invariably includes or invokes the aim of reducing somebody's uncertainty about something. The complementarity between investigation and intervention derives from the different starting points from which each attempts this reduction. If, in cybernetic terms, one wishes to steer the development of a system deliberately in some desired direction then the law of requisite variety (see Ashby 1956) suggests that the variety of controlling or directing responses of which one is capable must match the potential variety of the system of interest. Thus in a peak hour power cut, traffic chaos is highly likely if there are not as many policemen on point duty as there were traffic lights in operation before the cut occurred. If the policemen are available then uncertainty concerning traffic flow is reduced. However, in a very large city with a very low police budget, uncertainty can still be reduced by careful and conditional projective analysis. In other words the future uncertainty of free-flowing traffic generated by the capacity to intervene is replaced by the knowledge that if the lights go out traffic will be at a standstill within thirty minutes.

When it comes to a planning approach to the economic and demographic bases of a local or regional community we are in a position somewhat similar to that of the city fathers with a low police budget. The only way in which we can reduce our uncertainty is via projective analysis and so as a result such analysis takes on a very considerable significance. There may be certain ways in which we can bring some marginal influence to bear upon the population and unemployment levels of the areas for which we are responsible, but such policy instruments as are available fall, whether for good or evil, far short of the variety implicit in the systems of interest. Accordingly we are thrown back upon an almost total dependence upon investigation and prediction in pursuit of our low-level goal of uncertainty reduction. The only ways in which we can provide ourselves with an operating environment in which specific social policies and physical plans can be even sensibly discussed is through projective analysis of the economic and demographic contexts in which they must be implemented. Of course there are many other contexts which significantly affect the success potential of plans and policies, some of which will be discussed in subsequent chapters. However, none are quite so basic in an aggregate material sense, or quite so critical in defining the limits of local and regional political action, as those of economic and demographic structure.

How such fundamental aspects of local and regional structure are and should be handled analytically is, of course, an enormous question. The relationship between the two structures is possibly the critical issue which requires elucidation before sensible analyses can be performed, but it is by no means self-evidently clear that it makes sense to ask simple black and white questions as to the primacy of one or the other. Certainly most studies of economic or demographic structure typically make implicit or explicit assumptions about the relationship in order to facilitate tractable and specific analyses. Moreover these assumptions generally amount to something like the following two propositions:

(i)   aspects of either economic or demographic structure can at least be described analytically without recourse to specific and simultaneous treatment of both; and

(ii) in a state of relative ignorance as to the nature of the relationship, it is probably best hived off for separate treatment, that is as the independently visible phenomena of migration, unemployment and changes in the activity rate.

In order to expedite the discussion in the main sections of this chapter, we shall accept these assumptions, examine economic and demographic techniques of analysis separately, and reserve more general judgement upon relational as well as other matters for the conclusions.

Economic analysis

Techniques for the formal analysis of local, urban and regional economic analysis have received the attention of economic geographers and planners on and off now for not much more than twenty-five years. Since most current analysis in this sphere is dedicated by and large to exploring the behaviour of such crucial indicators of local welfare as income and employment, and since this has now become one of the most respectable fields of applied social research, it is surprising that its history is so short. Local political interest, at least in the latter of these two indicators, has a far more respectable pedigree. The explanation is in fact not hard to find. It lies in what, with the benefit of hindsight, we must regard as the incredible intellectual dominance of the classical and neo-classical economic models in the first half of this century. Economists simply refused to acknowledge systematic and enduring variation in regional wealth because the assumptions upon which their studies were based simply could not accommodate such a phenomenon. At best local pockets of higher than average unemployment were treated as the products of lags and bottlenecks, in the fullness of time to be automatically eradicated through rational entrepreneurial responses to market movements in the price of labour.

Myrdal (1957) and Hirschman (1958), working independently in the 1950s, were to effect a radical change in this position, and so to become responsible for an emerging academic interest in spatial economic issues. What they came up with, in effect, was the first disequilibrium model of regional growth. They argued that the decidedly limited advantages of a backward area would, under normal circumstances, be insufficient to offset the cumulating internal and external economies of a neighbouring expanding region. Thus the declining economy would be cumulatively drained both of premium labour and of raw materials, and welfare discrepancies would increase rather than disappear.

This pioneering work, and an enormous body of theoretical and empirical investigation since, has provided planners and policy-makers with such techniques of spatial economic analysis as they now possess. The questions that these techniques are intended to answer are essentially still those that derive directly from the original arguments of Myrdal and Hirschman. They are, first, is there a locally idiosyncratic competitive advantage or disadvantage observable in the development of a given economy, and second, if indeed there is one, in what feature or features of that economy (including external pressure) does it reside and what is its future impact likely to be?

Though it will be essential to look directly at examples at least of the techniques that are and might be invoked to provide answers to these questions, it is more important from the point of view of this chapter to be exactly clear about what these questions actually mean. This is especially the case as they are the sort of questions that are repeatedly asked in one form or another when aggregate analysis is expected to provide the answers. They appear to be related in a very straightforward fashion. Is there something locally which needs explaining? If so how do we explain it? In fact they suggest fundamentally different approaches to analysis. Though the second question can in principle contain an answer to the first, in practice this will only exceptionally be feasible. And a direct, straightforward and entirely satisfactory answer to the first question need never contain or even imply an answer to the second. The reason is that a question about the how much and when of a regional economy requires only, and most efficiently, some form of statistical or indicator analysis. However, a supplementary and related question about the wherefore and why requires a much more laborious form of structural analysis. And the link between the two may well be virtually non-existent. It may seem unnecessary to point out that we should not search both for the facts of local idiosyncrasy and for their underlying mechanisms with the same implements, but a problem does arise if we are not absolutely clear simply because we can go on to predict with either tool. The predictions will of course mean different things depending upon which sort of analysis has been performed. Whether or not either sort of prediction will mean very much is of course an even more basic question to which we shall return at a later point.

For the moment, the implications of performing each of these two alternative types of analysis in the domain of local economic uncertainty will be examined through a brief treatment of one of the most popular general techniques in each category. Shift and share analysis is a method designed primarily to answer the first of the questions noted above. In other words it is designed to tease out, as unambiguously as possible, genuinely local aspects of economic performance. The various forms of multiplier analysis, on the other hand, are designed primarily to address the second question about the mechanisms which underlie local performance, and are altogether more complicated instruments to operationalize.

Shift and share analysis: assumptions and method

The shift-share technique simply describes the development of a regional or urban economy as a local phenomenon by attempting to control for non-local factors which are assumed to influence that development. Thus if, as is usually the case, the index of interest is employment, then the purpose is to examine growth or decline of local employment as a relatively autonomous phenomenon, rather than just as a small-scale reflection of the performance of the national economy. Over any given time period, it is possible to specify what would have happened to employment in a particular local industry had it grown or declined at the overall national rate of change for all sectors. It is further possible to measure how that growth rate would have deviated from this national path as a result of the differential success, again nationally, of the particular industry in question. The difference which remains between what actually happened to the local industry and what might be accounted for in both of these ways, can therefore be interpreted as that which for one reason or another is uniquely local in terms of economic performance. In effect what the technique does is to compare actual performance with what would have been expected given national rates of change applied sector by sector to an economy starting out with a particular structure. If the economy is dominated by declining industries its expected performance, even on the basis of national rates of change, will be less than that of one characterized by fast growing industries. In such circumstances if it does turn out nevertheless to have exhibited an unexpectedly rapid expansion in employment over a given period, then there would appear to be no alternative but to search for some characteristic of the economy which was not shared by its neighbours. Of course this may well turn out to be a characteristic which is not in the long term best interpreted as 'local'. For instance, it may have been due to the inability of firms to expand in a neighbouring economy, or to the incentives offered by a nationally determined regional policy.

Shift and share analysis: limitations

In general one of the main practical difficulties with the technique is that it is not really feasible to differentiate so-called local from non-local factors by simply disaggregating employment into industry groups and applying adjustment factors based upon national data. Clearly the size of any local or regional idiosyncrasy which is described by the technique will be sensitive to the levels of both sectoral and spatial disaggregation. In the limit, as the sectoral disaggregation becomes finer to the point that each firm constitutes a unique sector of the economy, the idiosyncrasy (or differential shift as it is sometimes called) will disappear altogether. Likewise, as the size of the region increases to the point where the region and nation become coterminous, the idiosyncrasy will also, and necessarily, disappear. It would thus be dangerous to assume any absolute significance in a descriptive analysis effected through the use of the technique.

Apart from the relativity of results, perhaps its most important general limitation – for there are many other particular criticisms reviewed in the literature (see for example Houston 1967, MacKay 1968 and Randall 1973) – is that the technique isolates features of economic performance for explanation rather than offering any sort of explanation of its own. It is therefore far more appropriate for historical descriptive analysis than for projective use. For if it is used to predict changes in employment, it becomes nothing more than a very simple extrapolative device. Independent estimates of national growth rates by sector are required, but so are estimates of the residual local component of employment change. The crudest method, and that which has proved most reliable for short-term projections (see Brown 1969), simply freezes the most recently measured differential component for each sector, i.e. that appropriate to the preceding time period. In other words, the argument underlying a crude predictive use of the technique is that those aspects of local performance which are not explicable in terms of national growth rates are likely to be stable over time.

Shift and share analysis: applications

That this is a wildly heroic assumption hardly needs stating. How precisely we should respond to this vulnerability is something which needs much more careful thought. At one level, as suggested above, analysts have responded by avoiding predictive use of the technique altogether, but retaining it for historical description. Many regional and sub-regional authorities (see for instance Greater Manchester Planning Team 1975, Northern Region Strategy Team 1976 and West Central Scotland Plan Team 1974) have employed it for exploratory or diagnostic purposes. More recently, studies have addressed the inner city economic problems using the technique (see Rhodes and Moore 1973) and have made extensive use of the shift-share method to compare the employment histories of areas in receipt of regional incentives with those in which policy controls have been neutral or negative. The differential component is then interpreted as the effect of regional policy. Chalmers and Beckhelm (1976) have alternatively attempted to explain variations in the differential in terms of indicators of competitive advantage, such as market potential, using standard statistical procedures. And, most recently, Fothergill and Gudgin (1982) have based a systematic review of British manufacturing employment trends upon a comprehensive shift-share analysis in the course of assembling an eclectic account in terms of industrial structure, urban structure, plant size profiles and regional policy. In each case the basic statistical technique has been used as a starting point, simply to isolate a relatively unambiguous variable which can then be further analysed, or explained, in other ways. All further analysis, such as the comparisons of specially selected groups of areas (as in the case of Rhodes and Moore and Fothergill and Gudgin) or the statistical test of a theory of industrial location (as in the case of Chalmers and Beckhelm), must stand or fall on its own merits.

Export-base analysis: assumptions and method

In such circumstances, and given some of the more technical criticisms of the shift-share method, analysts have often opted to skip direct treatment of the first question as to the existence of local economic idiosyncrasy and move straight to the question of explicating and untangling such idiosyncrasy whether or not it exists. This is not as strange a ploy as it sounds, since whether or not over a given period of time cities and regions happen to exhibit differences in their economic success rates, there must be some sense in which each coheres as an independent as well as an interdependent economy. It is therefore worth investigating the mechanisms of this independence irrespective of the idiosyncrasy or similarity of their effects in the past, since if we are to achieve idiosyncrasy in the future (i.e. self-sustaining growth in a local or regional economy) it will have to be effected through the operation of these mechanisms.

   Multiplier techniques are widely used in one form or another to provide predictively useful summary descriptions of the mechanisms of a local economy. At least when used exclusively, they share a relatively modest view as to the extent of an economy's potential for self-sustaining growth. Whether the analyst uses the simplest export-base model or the most sophisticated regional input-output table, change in the economy is assumed to be triggered externally. In other words, however detailed the treatment of the structure of the local economy, all strict multiplier models are demand orientated in the sense that external demand for the goods and services of an area is said to initiate growth and internal demand to extend or 'multiply' it. These are the only positive change-inducing mechanisms included within the framework of multiplier analysis. Samuelson (1939) has of course shown how the multiplier and accelerator principles can be combined, so that a model results in which income is not only 'multiplied' through the operation of demand mechanisms but is also 'accelerated' endogenously through the lagged impact of income expansion upon investment. Such an extended model will describe a genuinely self-sustaining pattern of local growth, but since the accelerator principle is based upon a ridiculously crude behavioural premise, its use, at the local and regional level at least, is very rare.

The sheer modesty of the multiplier approaches is not, however, sufficient to guarantee their immunity from criticism. Their widespread use is by no means uniform. Thus each new application introduces a marginally or substantially different variation on the basic theme. The use of the input-output model at regional level can convincingly be interpreted as a response in part at least to the inadequacies of simpler multiplier approaches. The simplest of all is undoubtedly the export-base model. This attempts nothing but a precise description of the circulation of income which is generated in the short term through export sales. Nothing can be sold unless someone wants to buy it, and, in any economy, none can increase their demands unless they possess the disposable income with which to give voice to those demands. In the short run it is argued that new income will only come from increased purchases of local goods from outside the area. However, once the income has been generated some will accrue to local residents and some of this will be spent on locally produced goods and services. The final increase in income will thus be a multiple of the original export sales simply as a result of local economic interdependence. The tighter the interdependence, the longer the income will keep circulating and the bigger the multiplier. Moreover, since the income which local residents spend on local goods and services is the same as that which local firms receive for satisfying local needs, this multiplier, so the argument goes, can be calculated simply by means of an appropriate classification of firms rather than a detailed local expenditure survey. The classification can be approximated simply by converting to units of employment rather than income and comparing the size of each local sector – as a proportion of total employment – with the equivalent national ratio. To any extent that an area has more than its expected share of an industry, to that extent it must be exporting. A multiplier can be computed by relating the exporting, and thus employment-generating, fractions of each sector to the rest, which must by definition be servicing local needs and thus simply effecting expansion of employment through channelling local income.

Export base analysis: limitations

The employment approach to export-base analysis using location quotients, as described above, piles assumption upon assumption in an alarming fashion.

However, whatever the method for deriving the multiplier, when it comes to a projective measurement of the impact of a change in export demand, the most critical problem of any export-base study is the way in which very different aspects of economic behaviour are first confused and then collectively assumed constant. A constant ratio between local serving and export sectors means that:

(i)   local residents must continue to allocate expenditure between local and other providers of goods and services as they have done in the past,
(ii) they must continue to consume the same proportion of their income as in the past,
(iii) local firms must continue to export outputs and import materials and labour as they have in the past, and
(iv) they must continue the same pattern of production and interchange of intermediate goods and services within the local economy.

The problem, in other words, becomes almost exactly the same as the one faced by shift and share analysis if that is used for projective purposes. For, although the model described above attempts to operate by means of structural rather than statistical analysis, the structure isolated by the crude export-base multiplier lacks any real theoretical coherence. However, the multiplier is by definition related to a specified core attribute of the local economy, rather than an unspecified one. Thus all we can say about a local economy when the differential shift drops to zero is that we cannot yet discern any way in which it differs from the national economy (except that it is smaller). If however the export-base multiplier drops to zero, this means that in a very real sense the local economy has ceased to exist. Its composite units – whether individuals or firms – participate in economic relationships only with units outside the economy. The multiplier, in other words, captures at least the element of interdependence which characterizes an economy.

The problem is, however, that what appears as a theoretically precise and meaningful concept turns out to confuse at least four aspects of economic behaviour, undertaken by different actors and according to different social rules and interpretations. If our multiplier prediction turns out to be wrong we shall not be able to say whether this is due to changes in the patterns of consumption chosen by individuals or to changes in the patterns of production chosen by entrepreneurs. In this respect, therefore, our device is really no more helpful than an avowedly statistical technique which simply isolates features of economic change for explanation without pretending to provide an explanation at all. To assume that an export-base multiplier remains constant over time is thus in a sense an assumption built upon no firmer ground than any which involves extrapolating an observed but unexplained trend.

Export-base analysis: applications

The fragility of this assumption is increased if the multiplier is computed, as is generally the case, using the location quotient technique outlined above. For then the size of the multiplier becomes just as sensitive to the levels of spatial and sectoral disaggregation as does the size of the differential shift in shift-share analysis. Moreover, systematic biases will occur if productivity or consumption patterns vary from area to area or region to region.

Despite these problems, multiplier analysis is still being applied, particularly at a sub-regional scale. The technique has been used to explore the likely impacts of new manufacturing plants (Greig 1971) and key institutions in the tertiary sector (Brownrigg 1973 and Ashcroft and Swales 1982). More generally it has been used to explore the ramifications of demand generated through tourism (Archer 1976) and even as an indicator of the impact of industrial closures and rationalization (Brownrigg 1980). Most of these studies build upon the work of Tiebout (1962) and Weiss and Gooding (1970) though details of the method are varied.

Input-output analysis: assumptions and method

If disaggregation is a compensation for theoretical imprecision, then input-output analysis would seem to escape the trap of statistical obfuscation. Whilst export-base techniques typically make do with one or a small number of multipliers, input-output models generate a total which is more than the square of the number of sectors into which the economy is divided. They may best be thought of as analytical devices designed to overcome two of the most severe limitations of the export-base framework. First, and fairly obviously, the simpler model gives no indication of the differential impact of any multiple increase in income or employment upon the various sectors of the economy. Second, and following on from the above discussion, the model confuses at least two conceptually different multipliers – the Keynesian consumption multiplier and the Leontieff technological multiplier. Put simply, the input-output model attempts, via a sort of double entry accounting procedure, to monitor the sectoral impact of exogenous changes in demand, whilst controlling for the independence of the two multiplier mechanisms. Each sector is described as servicing a whole range of different demands. There are those imposed by local consumers, which depend upon local household consumption propensities, and those imposed by firms in other sectors, which depend upon their production functions and the overall level of integration in the economy. The technique simply converts the flows between each and every other sector into coefficients describing the input demanded per unit output. The assumption of constancy amongst these coefficients and the inversion of the matrix which contains them (to reflect second order and subsequent input requirements), permits a whole range of prediction, impact and multiplier analysis.

Thus if the demands imposed by local households are excluded from the matrix before it is inverted, the multipliers derived, and known as production multipliers, describe simply the direct and indirect effects of, say, a unit change in final demand for the products of one sector. Each of these will comprise the sum of the original unit shift in demand plus the necessary expansion in all sectors supplying the expanded sector, expansion in the sectors supplying the intermediate suppliers and so on. In other words these are the so-called technological multipliers, describing just the impact of local productive interdependence in the economy. If local households, as consumers and employees , are included within the main coefficients matrix, the resultant multipliers, known as income multipliers, will describe the effects induced by local consumption as well as the direct and indirect results of exogenous demand shifts. Not only will expansion result from structural interdependence, but it will be augmented as residents earn extra wages and salaries from the expanding sectors and consume proportions of those increments by purchasing from local firms.

The greater sophistication of the input-output model naturally means a greater variety of potential applications. It would be an incredibly laborious device for the sort of broad and semi-continuous historical description provided by the shift-share method. However, current structural analysis is one of its most fruitful areas of application. The intelligence provided by a base year input-output table for a city or a region enables a level of informed economic policy discussion which cannot be achieved in any other way. Each of the important links can be isolated individually, as can highly interdependent industrial and commercial complexes. The level of dependence upon single large employers of labour can be precisely quantified and the existence of significant and import-inducing gaps in the structure of the local economy may be systematically traced.

Inevitably, though, a mechanical device like the input-output model lends itself most readily to future orientated analyses.

Given that knowledge of the nature and health of the local economy is an essential input into any policy discussion, and that uncertainty about these crucial economic factors can only be reduced through analysis of one sort or another, the input-output technique has invariably been applied in one or other of its several projective modes whenever it has been calibrated for a region or smaller local area. Multiplier techniques are almost intrinsically predictive in orientation since the multiplier itself simply converts a current event into an eventual outcome.

Of course the ways in which the multipliers may be collectively and individually used to reduce economic uncertainty are several. The so-called 'neutral' forecast of the overall effect due to an expected change in exogenous final demand for the products of all local sectors is only one way of predictively applying the technique. The advantage of the device lies in the detail characteristic of its output. Thus production targets may be set, sector by sector, based upon some desired level of final demand. Individual rows and columns can be added to or subtracted from the matrix to permit the measurement of impact effects. The introduction to or withdrawal from the local economy of a single large productive unit could, for instance, be readily converted into a total income or employment effect by such an extension. And the addition of further 'non-economic' rows and columns, to produce the so-called 'dog-leg' matrix, will in principle permit almost indefinite elaboration of any expected disruption emanating from a hypothesized change in the processing core of the economy.

Input-output analysis: limitations

The important point from the point of view of this discussion is that each of these applications of the technique depend upon the constancy of the input coefficients over the prediction or impact assessment period. In other words, as output in each sector changes, up or down, so each of the input requirements change in a strictly linear fashion. Economists have voiced many reservations about this assumption at a purely technical level. They have argued the importance of non-linearities such as scale economies, the potential for capital/labour substitution, inventory accumulation and stock depletion. What is even more worrying at the regional and urban levels is that these coefficients are not purely technical. They are best interpreted as trade coefficients since any producer can vary at will any particular input coefficient without modifying the technical profile of his production lines simply by shifting from an intermediate supplier one mile within the region to an equivalent one a mile outside. One of the incontrovertible trends of the past thirty years has been a continued opening up of local and regional economies. The modern multi-national corporation now makes even national boundaries look arbitrary for the purposes of economic analysis. And so as technical coefficients become trade coefficients and trading patterns increase in flexibility, the predictive reliability of the input-output technique at regional and sub-regional scales diminishes.

Even its apparently substantial theoretical advantages over the simple export-base method – bought at a substantial price in terms of data requirements - begin to disappear when it is used projectively. For descriptive purposes, and assuming that all the appropriate information has been collected, it is indubitably true that the input-output model separates processing multipliers from consumption multipliers and is capable of generating a vast array of each sort. However, in performing an impact or forecasting analysis, the model assumes that a household exhibits a sort of production function which behaves exactly as does that of any firm in the economy. In other words it remains constant. The production of a population's desired lifestyle requires the maintenance of a constant aggregate ratio between all purchases made within the regional economy, just as the production of a sector's output requires the maintenance of a constant aggregate ratio between all its inputs. Though the consumption propensities and trade or technical relationships are measured separately in an input-output model, rather than being lumped together as in the export-base formulation, the assumption of across the board constancy means that in practice nothing is gained but a greater amount of predictive detail. It has in fact been shown (see Billings 1969 and Garnick 1970) that the two sorts of multiplier are algebraically identical. The input-output formulation simply represents a disaggregation of the export-base model. Thus given similar inputs the aggregated output of a predictive use of both models will be the same.

Moreover, a post mortem on the basis of a fully documented failure of each technique will not necessarily be made any easier by the greater descriptive detail of the input-output model even if we know precisely which sectors were predicted wrongly and by how much. This is because any error may arise in a theoretically infinite number of different ways. A rich structural description, which an input-output table can undoubtedly provide, only goes on to provide a structurally transparent prediction, forecast or impact assessment if it also says something explicit about the processes of structural development. The assumption of constant coefficients means that nothing is said about these processes. Thus unless everything external to the economic mechanism responds passively and instantaneously to maintain its stability, no sort of future orientated analysis will take us beyond the original description.

In other words we learn nothing about the development of structural or trade links in the economy simply by monitoring the predictive performance of an input-output model. A device which offers a sort of structural description of the local or regional economy therefore becomes merely another analytical black-box for predictive purposes, fundamentally no different from a shift and share or export-base technique. If we are to learn about the dynamics of trade coefficients we must first learn about the dynamics of production functions and then maybe incorporate such learning within the essentially static framework of the input-output model. Of course economists have been grappling with the limitations implicit in the static formulation of the model for years. Broadly speaking they have adopted one or another of two alternative strategies. Several have examined explicitly the stability – or instability – of technical and trade coefficients over time (see Moses 1955 and Riefler and Tiebout 1970). The conclusion most commonly drawn is a fairly obvious one. Trade coefficients are observed to be unstable over time. The equally obvious response is to attempt to dynamize the model (see Leontieff 1966 and Miernyk 1970), to introduce exogenously a process of change into the matrix of coefficients so that a structurally interesting static model becomes also structurally sensible for predictive purposes. The alternative of simply accepting the paradox is never even countenanced as intellectually respectable even though it is what inevitably happens time and time again in practice. Regional and sub-regional input-output models are devised in one way or another and the information they offer about the current structure of the local economy is gratefully received. Impact analyses and predictions are then (sometimes apologetically) performed, and whilst academic economists continue to grumble about the data constraints which preclude the calibration of dynamic models, those of a more realistic disposition get what they can from straightforwardly static examples of the genre.

Input-output analysis: applications

In recent years there have been many applications of the basic model at regional or sub-regional scale. In the UK, most of these applications have been produced either directly by planning practitioners with a view to examining the structure of the economy for which they are responsible, or as the result of a collaborative exercise with an equally practical end in view. The Cleveland study (Cleveland County Planning Team 1975) was designed specifically to explore the impact of North Sea oil discoveries upon the sub-regional economy. Studies subsequently performed in Merseyside (De Kanter and Morrison 1978), the Northern region (Northern Region Strategy Team 1976) and Shetland (McNicholl and Walker 1979) had more general exploratory and impact analysis purposes but the connection with policy issues was in each case apparent. And in each case the description provided by the transactions matrix was regarded – for all its deficiencies – as of crucial importance in the use of the technique.

Practitioners, in other words, rarely get to the point of worrying about the asymmetry between the structural sophistication of the model's description and the statistical simplicity of its use as an impact testing or forecasting device. The epistemological argument that we shall propose is that they are probably adopting the most sensible strategy since the paradox is endemic. The most interesting and useful of aggregate structural descriptions will always become a black-box when applied predictively. To attempt to dynamize it is thus an exercise which is of limited value. Formal statements about the future, through the conditional extrapolation of a frozen past and present, are in a fundamental sense the best we can achieve. We should therefore invest more attention in improving the ways in which we interpret such statements, and less in a fruitless quest for a level of quantitative confidence with respect to the future that will always and inevitably elude us.

Demographic analysis

The force of this argument is probably most clearly demonstrated by examining the aggregate techniques available to the planner for demographic analysis and projection. The similarity between the types of method typically used for population description and forecasting and those, just discussed, for the treatment of economic factors is truly remarkable – so-called structural devices being clearly distinguished from cruder statistical approaches. In demography, however, each category has a somewhat longer history, presumably because of the consistent availability of local census data in most countries for many years now. Moreover, the development of each can be readily traced to attempts at answering just one major question, rather than two as in the case of economic techniques.

In local demography the emphasis is almost invariably upon prediction. The current and recent structure of the population base is of course always of interest but is rarely regarded as problematic in quite the same way as in the case of the economy. Policy-makers and planners believe that they know quite unequivocally what is meant by population structure. It has to do with really basic typological dimensions such as those of age, sex and family or household structure. Occupational and industrial classifications seem very arbitrary by comparison. Moreover a comparatively recent age and sex breakdown of the population is almost always readily available, no matter how small or large the spatial scale. Not only are local employment breakdowns much harder to get hold of in many countries, but even when available they are rarely regarded (witness the techniques discussed in the last section) as adequate representations of the local economy's structure. Structure in the economic sphere much more obviously extends to include the notion of interdependence, with all that that entails in terms of analytical complexity.

In other words, population analysts can afford to speculate more about the future since a description of the current situation appears to be an analytically simpler problem. It is also, however, probable that this single-minded interest in prediction reflects a greater pessimism about other means of reducing uncertainty. Whatever local policy-makers feel about the intractability of the problems of the local economy, the view of that economy as at least in principle capable of management is still strongly felt. It is after all not long since economists believed local economies to be self-regulating. That this position has been undermined (for some at least) only means that we must now accept the importance of external regulatory aids in the form of spatially differentiated monetary, fiscal and employment policies. The principle that control is theoretically possible is rarely questioned. The demographic base, or at least its fundamental mechanisms, are quite differently regarded. Though planners and local policy-makers often assume that they can exercise a certain influence over migration, the processes of natural increase or decline are considered for all operational and analytical purposes to be beyond the scope of local control. The only way, in other words, for the planner to reduce future uncertainty in this sphere is to produce reliable forecasts.

Trend analysis: assumptions and method

As indicated above, however, two conceptually distinct techniques have been employed. The simple statistical devices, like employment shift-share techniques, attempt no mechanical (or any other) explanation of the local or regional idiosyncrasies they isolate and describe. Since, however, demographic statistical techniques have a single projective rationale, the extrapolation of these locally idiosyncratic patterns becomes all important. It therefore becomes critical to find some relatively stable relationship between aggregate population change and some other variable whose future distribution is either certain or highly predictable. The only variable which unambiguously satisfies the second of these conditions is the passage of time itself. Thus if population change in the recent past demonstrates a pattern of strictly linear or smoothly curvilinear movement, it may be extrapolated either graphically or statistically without much difficulty. Alternatively, if a stable relationship has been observed for some time between a study area (usually a city or region) and some pattern area (such as the nation) and a reliable forecast exists for the pattern area, then a simple ratio projection can readily be made for the study area, again by means of simple statistical or graphical comparison. In neither of these circumstances would an analyst claim that a causal mechanism had been isolated and described.

The assumption in the first case is simply that some sort of aggregate momentum builds up. Whatever the causal mechanisms are, their historical coalescence produced such a stable pattern of aggregate movement that it seems reasonable to postulate the continuation of this coalition of forces for at least a short period into the future. In the second case the argument is simply extended. If we observe a parallel between the momentum in our study area and that in some pattern area, and we have good reason to believe the momentum will be broken in the pattern area, then we may reasonably postulate the continuation of the parallelism and predict a discontinuous shift in population trends in the study area (see Isard *et al.* 1960 or Pressat 1978 for fuller discussions).

Trend analysis: limitations

There are a multitude of technical difficulties to be overcome in the implementation of analyses of this sort. Planners and demographers have invested a great deal of time and energy in testing marginal variations in functional forms, alternative logarithmic transformations, variously lagged versions of the ratio function and so on. However, the major difficulties are to do with the highly aggregated nature of the model's output and the extreme fragility of its central assumption. The latter boils down to a belief in the force of aggregate inertia. Since there can be no direct causal connection between population changes and either the passage of time (as in simple trend formulations) or the process of spatial aggregation (as in ratio techniques), the analyst is left not with a theory but with an article of faith – the belief being that whatever the mechanisms are, their apparent aggregate stability in the past will persist into the future. The social and biological processes involved, and the patterns of their spatial arrangement, are so deeply ingrained that their future trajectories are unlikely to be very different from those of the recent past.

It hardly needs stating that this is a most inadequate theoretical base upon which to build an approach to forecasting. Fertility rates, migration rates and even mortality rates may all change over time and so precipitate a shift in the historic trends. Moreover, even if these rates were all to remain constant functions of the fundamental age and sex structure of the population, that structure might itself conceal a pattern of movement in the most probable future levels of aggregate population which may not be revealed simply by examining past trends and ratios.

Trend analysis: applications

Despite this theoretical *naïveté*, simple extrapolative devices have been used extensively for population forecasting, and examples can be found at all spatial scales. However, in recent years, dissatisfaction, both with the crudity of the output and with the theoretical limitations of the assumptions, has occasioned a shift away from what was at one time an almost universal reliance upon simple statistical devices of this sort. The most recently completed review of analysis strategies in UK structure planning indicated that in Wales only three out of ten counties relied exclusively upon trend techniques for population forecasting (see Bracken and Hume 1981). An earlier review (see Barras and Broadbent 1979, 1981) also indicated a shift towards more sophisticated and disaggregated methods of population analysis and forecasting.

In other words, there has, for some years now, been a considerable demand from planning analysts for more elaborate and reliable demographic techniques. It is therefore no more surprising that regional demographers have shown such a great interest in direct structural analysis of local populations than it is that regional economists have invested so much effort in techniques such as input-output. The parallels even extend to the choice of procedures. Just as economists have attempted to isolate mechanisms which describe analytically the employment or income growth generating and sustaining potential of the local economy, so demographers have attempted to isolate analogous multipliers for a given local population structure. The chief difference between the two parallel efforts resides simply in the fact that, if migration is ignored, the concept of structure in demographic analysis is not one which carries with it the connotation of aggregate integration and interdependence. The economy is no more than an artefact, built fundamentally upon the social principle of exchange. If transactions do not occur the economy ceases to exist. The population base, in contrast, is characterized by a far higher level of disaggregate autonomy because the fundamental laws are biological, and social rules and practices can only operate within those frameworks.

Cohort survival analysis: assumptions and method

Practitioners, disenchanted by the inability of statistical techniques to capture even this crude element of structure, have opted more and more for approaches which acknowledge at least the basic biological regularities. Assuming for the moment that there is no net migration into or out of a region, then the mechanisms which underpin aggregate growth and decline of populations are simply natural increase and mortality. The population will change over time simply as a result of the rates at which children are born into it and subsequently exit through mortality. Of course this simple micro level tautology which describes the mechanisms of population change is capable of more or less meaningful aggregation for analysis purposes. By far the most popular approach, known as the cohort survival method, involves simply disaggregating population into age and sex-specific categories, on the basis of the biologically reasonable premise that birth and death rates vary more significantly with this dual classification than with others. Social indicators such as class, income or education may also be related to variation in the critical rates, but not to the same extent. The simple cohort survival technique merely applies, through a matrix formulation, age and sex-specific survival rates to the population size in each appropriate category in order to estimate the numbers who will still be alive in the next and subsequent time periods. Age-specific birth rates are also applied to female population totals in the child-bearing categories to generate predictions of the expected natural increase in each time period. In this way a picture is built up of the likely development of the age and sex structure of the population in future time periods, as well, of course, as an indication of its likely absolute size at each point in time. Estimates such as these are naturally of considerable value, if reliable, for the purposes of planning in all functional areas and at all spatial scales (see Rees and Wilson 1977).

Again, though, the crucial question is a theoretical one. The argument that the transparent and straightforward structural description upon which the cohort method is based becomes little more than an opaque and statistical one for predictive purposes may at first sight seem somewhat unfair. However, the point can be brought home most forcefully by reintroducing the migration component of population change, and comparing a historically descriptive use of the technique with a prediction. One very useful application of the cohort method is for estimating net migration as a residual. If we know the population totals broken down by age and sex both at the beginning and at the end of some past time period and we know also what the birth and death rates (again by age and sex category) were over that period, it is of course possible using the cohort method to apply those rates to the initial population distribution in order to derive an estimate of population change resulting purely from natural increase and mortality. There will undoubtedly be measurement errors due to the averaging of the critical rates and possibly also due to the necessity of using rates originally calculated for a larger area than the region or city of interest. However, these should be acceptably small and so net migration, positive and negative, can be fairly reliably estimated for each age and sex category simply by subtraction.

If, as an extension, the analyst wishes to produce a prediction of the way in which population will change over a subsequent time period, all the necessary information is apparently now assembled. A survival matrix of birth and obverse mortality rates has already been produced, and, in addition, a migration matrix may be composed by calculating age and sex-specific net migration rates from the residuals derived above, or from direct migration measures if they are available. By applying each of these matrices to the vectors of current population and adding the results, estimates of the future age and sex profiles of the population may be readily produced. In effect the procedure assumes that all three sets of critical rates – birth, mortality and migration – remain constant from one time period to the next.

Cohort survival analysis: limitations

It is of course in this assumption that the problem resides, just as it did with the input-output model. In exactly the same way the rich detail of the prediction serves only to obscure the basic similarity between the cohort model and, say, a simple linear trend extrapolation. Though the procedure of calculating age and sex-specific birth, mortality and migration rates is of undoubted value as an index of the structure of the population over the recent past, a single period projection based simply upon freezing these rates is identical when aggregated to a crude growth composition analysis which involves applying crude aggregate birth, mortality and migration rates to single figure measures of population. It therefore permits no more in the way of effective post-mortems than does an input-output projection which is shown to be wrong. The fact of a wrongly predicted population size for a particular city or region may equally well be due to changes in any subset of the multitude of basic rates which collectively produce that total.

We may, none the less, wish to speculate about the source of the error, and be tempted to assume that it was in all probability due to change in the migration rates. This is because, in so far as the cohort model claims to capture aspects of population structure which are likely to be stably related to change mechanisms, the argument with respect to migration must be by far the weakest. The certainties underlying the cohort model are simply that eventually, after a process of ageing, we shall all die and that only women within a relatively narrow age band can give birth to children. By contrast there is really no respectable reason for assuming any sort of stable relationship between net migration rates and the age/sex structure of a resident population. Thus there is a fairly respectable, if rather negative, reason for assuming an erroneous demographic projection to be due to instability amongst age and sex-specific migration rates.

The important point, however, is that there is nothing intrinsic to the cohort model which enables us to perform such diagnostics. For predictive purposes the model obscures the critical relationships. Its ability to provide useful descriptions of population structure may be considerable, but its ability to provide more than statistical extrapolations when used for forecasting is zero. As in the case of input-output analysis this is because it says nothing about the dynamics of the change producing relationships it isolates and describes. The when, the why and the how of changes in birth, death or migration rates are by definition questions whose answers lie completely outside the framework of the model itself. Migration is primarily a spatial phenomenon, and so to treat its variation as analytically related purely to the age and sex structure of a single population is patently absurd. And though this particular demographic way of structuring a population is far more relevant in the case of birth and death rates, there are many other factors which may independently explain their variation.

Cohort survival analysis: applications

These problems have not inhibited the widespread adoption of the technique as one of the most basic tools of strategic planning practice. Its popularity – adequately documented in the reviews already quoted (Bracken and Hume 1981 and Barras and Broadbent 1979) – derives both from its pertinence and its ease of use. In disaggregating population structure into age and sex cohorts, it provides estimates which lend themselves ideally to use in various areas of policy analysis – particularly those concerned with facility and service provision such as housing, health care, education, welfare, leisure and recreation. Such facilities and services tend often to be age related in their market or client orientation. Generally the model can be fitted without recourse to primary data collection exercises and may be used in a projective mode as easily as for demographic description.

Most practical applications do in fact focus upon the projective use of the technique. Though migration is normally dealt with (if at all) outside the framework of the model itself, this does mean that strategic planning studies invariably come to depend at least upon the constancy of local or regional birth and death rates. Each of their infrastructure and service provision plans comes to depend upon the accuracy of their future estimates of population structure. And in this case, even the concept of 'structure' begins to look precarious.

As noted above, the key difference between a space economy and its demographic base is that whereas the former is an aggregate structure, the latter is not. There is no sense in which the whole thing hangs together as a network of inter-related parts. A society or a community may be said to exhibit some collective properties, but a population may not. Therefore the statistical inadequacy of the extrapolative use of a technique like the cohort model is even clearer than it is when an input-output formalism is applied predictively to the regional economy. If an aggregate structure exists, then a technique which mirrors that structure may be expected to perform effectively as a predictive device. If no aggregate structure exists, then an aggregate model can be no more than a contingently useful statistical summary which cannot possibly provide anything but a trend extrapolation. Its utility as a description is irrelevant when it comes to assessing its role as an aid in forecasting.

This distinction is very important if one's purpose is to improve the forecasting technique by incorporating functions which permit the continuous modification of theoretical rates assumed constant in their simplest versions. For whereas economists could be forgiven for single-mindedly seeking structural sources of variation in input coefficients, demographers would be hard put to justify such exclusive searches in the case of migration, birth or even death rates. There are no doubt a whole variety of aggregate social, economic and cultural factors which collectively define the circumstances in which are set the individual choices and discrete events which compose these rates. But there can exist no aggregate overarching and interlinked structure since ultimately each of these family choices or natural events are independent rather than transactional in nature.

Epistemological problems

Thus it seems that the most popular and descriptively useful techniques for urban and regional economic and demographic analysis share the same limitation when used for any sort of future-orientated impact or projection analysis. The simple statistical techniques such as shift-share and curve fitting are, in this context, theoretically opaque in that they say nothing explicit about the change mechanisms of the local economic or population bases which they describe. Input-output and the cohort survival methods have, at least for descriptive purposes, a far stronger theoretical base since each isolates and measures critical mechanisms of change. However, a snapshot measurement of these mechanisms at some arbitrarily selected moment in time is as near as each technique gets to a theory of structural dynamics. Thus for forecasting purposes each becomes fundamentally indistinguishable from the simpler black-box devices. Each achieves a prediction by assuming constancy in all the critical rates which describe change mechanisms, and so neither is capable of saying anything about the processes of change themselves.

Economists and demographers have typically responded to this situation by searching for ways of explaining variation in the actual mechanisms of change which can be quantified and incorporated within the structural models. The futility of such a search is seen most clearly in the case of the cohort survival model, for as indicated above, the demographic base is not in any aggregate sense a strictly structural system at all. The motors which drive the thing for predictive purposes are aggregations of decisions and events which are in no way systematically dependent upon a related series of similar decisions occurring elsewhere in the population system. The most that analysts can hope to find, therefore, is a framework of aggregate constraints whose force with respect to demographic change mechanisms is historically measurable and whose future shape is exogenously predictable. The point, then, is not that the demographic base is not capable of aggregate analysis, nor that prediction *per se* is intrinsically futile. It is rather that sophisticated mechanistic structural models are not the way to set about such a task.

The same is true, though less obviously, of macro level economic analysis. An aggregate structure actually exists in this case, fixing and organizing the channels of growth and decline. Increased production cannot be sustained without increased demand. The constraints which influence change in the economy are thus internal and structural as well as external and independent (though of course the latter exist also, and the demographic base is one of the most important of these). A technique, like input-output, which describes the economy's pattern of structural interdependence seems, therefore, to offer a significant prima facie advantage when it comes to the task of future-orientated analysis. The internal structure must persist or the economy will cease to function. In fact, of course, theoretically the structure need not persist at all. It certainly need not, as for predictive purposes input-output techniques assume it must, persist unaltered. For, just as in the case of the population base, the mechanisms of change are aggregations of the decisions of individuals. These decisions have to be negotiated with many others in a way which is not so necessary in certain other non-economic spheres of human activity, but this does not render them any more susceptible of aggregate structural prognosis. Any attempt to dynamize the input coefficients by adding to the structural sophistication of the model will fail because ultimately the process of negotiating business decisions is no more capable of structural analysis than is that of negotiating a contract of marriage.

The central epistemological difficulty is now quite clear. An applied human science defines principles of investigation which must be adhered to at all spatial scales and in the use of all analytical techniques. We cannot accept the essential similarity of subject and object in social research and at the same time apply aggregate economic and demographic techniques in a fashion which denies its relevance beyond some point along the micro-macro continuum. Change occurs in the structure of an economy because the individuals whose negotiations constitute that structure are innovative. They find new sources and types of raw materials. They devise new processes for producing goods and providing services. Similarly change occurs in the aggregate pattern of population growth and decline because individuals and groups self-consciously adopt new cultural standards with respect to family and working life. These standards affect the decisions they make about marriage and child rearing. We shall address the problems of analysing these sorts of decisions directly in later chapters. For the moment it is more important to explore the specific implications of reaffirming such epistemological principles for the sorts of aggregate analysis discussed in this chapter. Is there anything that can be salvaged from the traditional approaches to the economic and demographic bases? In what new directions – if any – should the further development of aggregate analysis techniques be pursued?

Synthesis

It cannot be emphasized too strongly that it is necessary to salvage something at this level. If planning has anything to do with the reduction of uncertainty then at the fundamental levels of aggregate economic and demographic structure the likelihood of achieving anything is dependent upon the ability of appropriately used analysis techniques to provide relevant and up-to-date descriptions as well as modestly reliable predictions or future-orientated impact statements. As far as descriptions are concerned, it has been shown that the input-output model, and to a lesser extent the cohort survival technique, are of considerable value. The former provides one of the most useful ways of approaching the issue of economic structure, and the latter a quick and convenient way of exploring population profiles and of deriving reliable estimates of recent and current net migration levels. Moreover, their potential at the level of historical and current descriptive analysis has rarely been tapped in practice since the emphasis is almost invariably upon projection of one form or another. The migration estimating use of the cohort method, described already, is rarely emphasized in demographic analysis whose ultimate purpose is some long-term forecast. Yet the importance of knowing something about the size and age structure of recent net movements in and out of a given area cannot be over-emphasized. If, moreover, input-output tables are composed in the only reliable way – i.e. by means of a survey – then their descriptive possibilities are almost limitless, for it becomes feasible to try all sorts of different typologies. Typically, whatever the data base, input-output tables are assembled once and for all according to strictly traditional rules defining the classification of the units of the economy. This is a great shame because a whole variety of different classifications can help illuminate the patterns of structural interdependence within an economy. The public sector can be finely disaggregated, for instance, or firms can be classified in terms of their labour intensity. Since at a regional or local level input coefficients can reflect only a mixture of technological and trade relations, there is no good reason for sticking with just one typology on the grounds that it is based upon technological differentiation. Far better to devise an economic analysis strategy which is flexible on this point.

When it comes to future-orientated analysis, the problems are of course very much greater. The first requirement is that any predictive analysis take cognizance of the implications of its status as a project in human science. What this means, as already indicated, is that its essentially statistical limitations must be acknowledged. The analyst must not be deceived by the apparent structural sophistication of a descriptive device since for predictive purposes it must inevitably become theoretically indistinguishable from the simplest of statistical techniques. What this means is that it must be treated with the same degree of caution. It is neither desirable nor legitimate to assume the possibility of dramatically reducing future uncertainty through the use of any sort of analytical technique, no matter how sophisticated the structural description upon which it is based.

There are, however, ways of setting up an aggregate projective analysis which will at least reduce the chances of simple-minded deterministic irrelevance. The three most important are, first, the employment of an integrated parallel forecasting style; second, the explicit treatment of the state as an interdependent, constraining and dampening set of agencies; third, the fastidious use of statistical risk analysis. The last of these is in a sense the easiest to explain. There are clearly a multitude of ways in which the uncertainties inherent in performing projective aggregate analysis can never be eliminated. However, that is not a good reason either for dropping such analysis altogether or for undertaking it without acknowledging the risk of error. The first method for getting to grips with the risks inherent in attempting future-orientated aggregate analysis thus involves simply being explicit about how large they may be expected to be. In other words the risks are not magically eliminated, but are subjectively quantified. Techniques variously called sensitivity checking and robustness analysis (see Rosenhead and Gupta 1968) form parts of what is generally meant by risk analysis. Of course, just as the future itself cannot be precisely predicted, so our degree of uncertainty about it must be equally uncertain. Yet we are normally willing to assess the risks associated with any forecast when, for instance, playing cards or betting upon horses. We are for instance likely to be much more cautious when playing cards with professionals. We require better odds if a race is six weeks hence rather than tomorrow. And of course the facts of the situation are relevant as well. We always bet more on a royal flush than a pair of nines, and more on a thoroughbred than a carthorse. In other words we combine our knowledge of the existing situation with our subjective beliefs about the future in order to make a decision.

The Bayesian approach to statistical decision theory (see for example Ackoff and Sasieni 1968) is a means of formalizing this synthesis. It enables the decision-maker to describe a probability distribution about a forecast by combining information about current uncertainty (to do with measurement error, sampling, aggregation, etc.) with subjective probabilities assigned to various alternative future states. Subjective probabilities are of course all that we can adduce to describe forecasting error that will result from the inherent uncertainty associated with the future. However, they are widely employed in private sector planning. In the public sector, however, such is the crudity of most forecasting exercises that not even hard information about current uncertainty is used to describe the error expectation associated with a forecast. It would be possible, for instance, with an input-output table derived by means of a survey, to describe the current variance associated with each input coefficient by comparing input profiles of individual firms within each sector. Subjective probabilities might then be assigned to the prospects for sectoral efficiency changes, shifts in inflation rates and movements in the level of final demand. Finally an expected local income or employment probability distribution could be derived by means of a sort of Monte Carlo technique – repeatedly running the model and at each stage varying its inputs in accordance with the current and subjective future probability curves. A similar approach could be adopted for use of the cohort survival model in predictive analysis.

The second method whereby the use of aggregate techniques may be improved is through greater interpretative integration. What I mean by an integrated parallel forecasting style is one which attempts to get away from the traditional planning penchant for treating functional categories of urban life as autonomous. Barras and Broadbent (1979) recently demonstrated the seriousness of this problem in a review of published UK structure plans. Chapters on employment, population, housing, retailing and recreation were rarely found to be as tightly interrelated as they might have been. In some cases different demographic forecasts were to be found in different sections of the same reports.

It makes no sense, however, to swing to the other extreme

and attempt to force the whole planning analysis exercise into the straitjacket of some horrendous comprehensive macro model. Physicists do not attempt to model the totality of physical systems and there is no reason to assume that planners will achieve anything more than the justifiable charge of arrogant irrelevance if they do not also side with the scientific angels in this respect. Advance in the social and human sciences is just as likely to be achieved through modest attacks upon small problems as it is in the more mature and successful natural and physical sciences. This does not, however, mean that the small problem of how other problems relate one to another is unworthy of intellectual effort. Such effort is in fact one of the few approaches available to us for partially checking the consistency of our forecasts. At the very least we are forced to consider the implications of inconsistency. If we make conditional and partial forecasts using input-output and cohort models as quasi-statistical devices, then imbalance between projected employment and population totals (measured in terms of current activity rates) must be scrutinized most closely, since it means that one or more of the assumptions most critical for planning purposes are wrong.

The input-output model assumes elastic labour supply, and the cohort model describes (in its least dubious form) only the results of natural increase and mortality. Thus an imbalance in one direction may signify short-term labour bottlenecks, which might prematurely terminate a phase of economic growth, and longer-term net immigration which in its turn might impose severe stress upon housing, education and health services. An imbalance in the opposite direction may indicate either inefficient labour hoarding or socially and individually painful unemployment, leading eventually perhaps to out-migration, physical dereliction and long-term damage to the social and economic viability of a city or region. The box in figure 3.1 below labelled migration/unemployment, etc., indicates our ignorance with respect to these crucial aggregate safety valves, but the fact that it is drawn as just one box demonstrates that these valves are not autonomous. They serve to integrate the economic and demographic bases of an area, and thus also to render illegitimate analysis programmes which rigidly separate them (see Breheny and Roberts 1978).

It is equally unrealistic to ignore the state. One of the

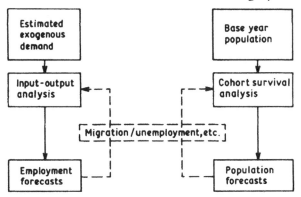

3.1 Integrated aggregate analysis

paradoxes of public sector planning is that while policy-makers are furiously dreaming up more and more ways of establishing greater interdependence between the state and the individual, policy analysts are pretending that such interdependence does not exist in their dogged reluctance to abandon neo-classical economic models. The argument that what is required is an indication of how people and economies would behave were they not constrained by state activity – a sort of hypothetical revealed preference analysis – makes no sense if there exists no *laissez-faire* base line upon which to build the analysis. Of course the cohort model is not economic and the input-output technique is as applicable to a situation of total state management as it is to that of the US economy of the late 1940s for which it was originally fashioned. However, both can and have been applied within an analysis framework which tacitly accords the state a purely marginal role. Thus the state can be and has been treated as just one more processing sector in the economy, passively and elastically responding to demands for its services. Similarly one can imagine a situation in which a very high level of infant mortality in a poor region might be extrapolated forward in a simple-minded fashion without questioning the possible impact of a determined state response to this situation.

A self-defeating attempt to overcome this deficiency would be one in which planning analysts attempted to describe and predict state action using language and devices similar to those employed in other areas of analysis. In his adoption of a cross-impact framework this appears to be a fairly close approximation to the approach that Bracken (1982) is advocating. If the above argument is accepted, it becomes impossible to predict, using mechanical techniques, the future behaviour of even the most highly structured of economic aggregates, let alone the complexities of public sector involvement in these structures. If for no other reason, the close proximity of the planner as analyst to what would count as his or her subject matter must render dispassionate commentary and projection quite impossible. Far better to capitalize upon this involvement. The products of the various parallel economic and demographic projections and impact analyses can then be monitored by the analyst as a conscious participant in the social processes they describe. The informal and formal consultations and discussions which generate policy thus become parts of the analysis process. The planner sets up a dialogue with the more mechanical devices employed so that aggregate analysis is fed by, as well as feeding, the processes of policy-making. The ways in which the state is involved in the past, present and future processes of urban life then become ways of reducing our uncertainty. Figure 3.1 can then be redrawn (see figure 3.2 below) to indicate how the private sector safety valves are complemented or attacked by state action designed to ensure stability and balance between the population and economic systems.

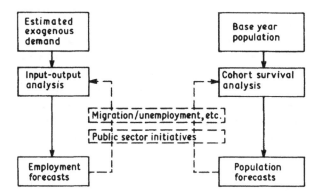

3.2   Integrated aggregate analysis incorporating the state

Each of the strategies described above represents a way of approaching the outputs of aggregate demographic and economic analyses when used in parallel for generating forecasts, enabling us to cross check, circumscribe and at one level to integrate the outputs of each model. In the near future it may, moreover, prove possible to formalize some of these strategies by building upon the techniques currently under development in the field of artificial intelligence. We shall explore some of these possibilities briefly in the next chapter. For the moment, however, it is worth noting that figures 3.1 and 3.2 indicate, amongst other things, the dangers of all forms of systems analytic representation. Though it is undoubtedly true that a mismatch between employment and population levels can easily lead to unemployment or migration, and though it is equally true that the state via its consumption, investment and other activities will attempt to intervene in these processes, the suggestion of simple negative feedback loops is not what is intended by these figures at all. It is very important to remember what sort of analytical process is represented by the major forward sequences in each of these diagrams. We have shown above how the input-output and cohort models both take on the properties of simple statistical devices when used for forecasting. They say nothing explicit about the dynamics of the local economy or population system. This means that any successes achieved result from the constancy over time of the production functions, fertility rates and a multitude of other relevant change mechanisms – or at least the occurrence of no more than random variations at the micro level which appear as constant coefficients aggregated at the macro level.

One of the purposes of deliberate state intervention, however, is precisely that of influencing these micro level processes in a decidedly non-random fashion. Monetary and fiscal policies are employed systematically to shift production functions through the manipulation of factor prices. Health and welfare policies may be designed in part to affect fertility and mortality rates. And even if we make the absurd assumption that the state remains passive, the so-called 'safety-valve' processes of under-employment and migration, must inevitably affect the dynamics of the local demographic and economic bases. The point is that since the forward models are static they are capable only of extrapolating existing trends. They are quite incapable of accommodating feedback loops which represent manipulations of the dynamics of the setting. They depend for their predictive success upon those feedback loops either not existing or operating only as they have operated in the immediate past. Since the former condition is a patent nonsense, this means that all the various feedbacks must continue to apply exactly the same dampening effects as they do at present if the models are to remain accurate.

The fact that they will of course not do so should not encourage us simply to give up any attempt at reducing our uncertainty as to the future of the local economy and its demographic context. It should rather encourage us to maintain an absolutely clear distinction between our modes of analysis. Thus forward aggregate extrapolations become useful not because they indicate what will happen but rather because they represent one sort of starting point for a quite different process of analysis which we may call dialectical interpretation. They constitute not a package of information but an intellectual challenge. And the point about such challenges is that they can only be met by thinking creatively upon the behavioural dynamics and structural inconsistencies of the situation. To respond to such extrapolations by multiple repetitions of the aggregate statistical model, each plugged with alternative input assumptions designed to reflect alternative public sector responses to the initial scenario, would be to misconceive the status of the techniques and to shrink from the responsibilities of the social analyst.

What is required is rather an approach which contrasts formal and mechanical analysis with approaches to theorizing which can never be formalized in any mechanical fashion. We shall return to the treatment of these informal techniques in later chapters. Here it is finally worth emphasizing the point that none of these stratagems should be thought of as an answer to the problem of intrinsic uncertainty in aggregate economic or demographic forecasting. The uncertainty derives from the ultimate partiality of any aggregate mechanical view of population and economic systems, not from any particular weakness of the popular techniques discussed in this chapter. The biological inevitabilities of the population system and the structural interdependencies of the economy merely define frameworks within which human choice acts as the sole and essentially unpredictable dynamic. Change occurs because creative individuals impose meanings upon the transactions in which they participate which then influence the content of that participation. The meanings themselves are established through a complex and partially deliberated process of cultural adaptation. And as part of this process, these meanings are consciously adjusted to a variety of circumstantial factors which include any predictions of which individuals may be aware. The self-defeating prophecy is thus part of the planning analyst's way of life – a part which must sooner or later be addressed directly.

4

Spatial form

Before we shift our attention to focus directly upon the motives and choices which either perpetuate or redirect the economic and demographic regularities discussed in the preceding chapter, it is necessary to pause briefly and examine the problems of one special type of elaboration to such aggregate analysis. The elaboration referred to is that which renders the analysis spatially specific. Just as the analyst attempts to reduce our uncertainty about the present and future of an area's economic and demographic bases, so he or she must sooner or later step down to the spatial and physical corollaries of that level of analysis since all activities consume space and many involve movement across it. The traversing and consumption of space involve complex human processes and generate durable physical products. Inevitably the state becomes heavily involved. Currently, central and local governments in all advanced political economies engage in a vast range of activities which both depend upon and influence the topographies of cities and regions. The policies they devise and the programmes they implement either involve direct physical implications – through the creation of social capital or the modification of spatial controls and incentives – or operate indirectly by influencing the human activities the pursuance of which implies the consumption of physical space. Thus a change in the structure of income tax through an increase in child allowances may well come to mean an increased demand for nursery education facilities in those areas of the country in which there are high proportions of young married couples. And the direct spatial involvement comes when and if the state accepts the responsibility for providing the facilities.

The extent of direct and indirect state involvement in the determination of spatial form does not render the task of the analyst that much easier. In a mixed economy, the problem, if anything, gets worse since public sector decisions are taken either with no regard to their spatial implications or in an attempt to strike some not very clearly specified balance between accommodating and deflecting market forces – in a situation in which those forces are themselves probably misconceived with respect to their spatial effects.

The response of the analyst traditionally takes one of two alternative forms. Location theory has in the past usually amounted to an attempt to describe the spatial outcomes of the concrete decisions of individuals acting as the agents of physically mobile operations in manufacturing or tertiary industries; as householders searching for residential accommodation; as commuters selecting a mode of travel to work; as shoppers choosing amongst retail outlets; or even as public sector officials responsible for applying spatial controls or distributing social capital. The issue in each case traditionally boils down to a debate about the behavioural – or (more recently) political – assumptions upon which sometimes extremely elegant deductive edifices are built. This micro or middle level approach will be treated in subsequent chapters. The alternative is to treat spatial form directly as some sort of aggregate phenomenon which may be described, quantified and possibly even extrapolated into the future. The purpose of this chapter is to review aggregate spatial analysis techniques which fall most naturally into this second category. Inevitably techniques which focus upon the aggregate profiles of spatial activity have little to say directly about the human processes which generate these profiles. Yet it seems equally inevitable that such techniques are bound to figure significantly in any programme of applied analysis. As suggested at the outset of this chapter the informed design and implementation of planning or other urban social policies is just as likely to be initiated by uncertainties which concern spatial form as it is by lack of basic economic and demographic data and understanding.

For this reason, as in the preceding chapter, we shall apply the principles of an applied human science retrospectively to the techniques of macro spatial analysis. We shall first attempt to clarify the sorts of uncertainty to which they are addressed, and then assess them against yardsticks which derive from the much broader epistemological frame of reference described in chapter 2.

At first glance, the most obvious way of classifying the techniques of spatial analysis must appear to be through some sort of detailed functional typology. Thus the spatial analysis of employment would be treated quite independently of the equivalents for, say, retailing or residence. In practice, at the aggregate level, techniques for spatial analysis fall much more conveniently into a much simpler two way classification based upon their overall strategies. Those approaches designed essentially for the summary description of map pattern are clearly demarcated from those intended primarily for the analysis of spatial interaction. It is this classification which we shall adopt, at least for the early stages of the following discussion, since it is the one which highlights most poignantly the problems of aggregate spatial analysis. The fact that there is a spatial dimension to the planner's role as a policy-maker does not necessarily mean that there should be a strictly analogous spatial emphasis within a coherently relevant programme of applied human science. The two categories mentioned above, of map pattern and interaction analysis, more or less exactly correspond to areas of research which, whilst focusing at the macro level upon the same spatial forms and processes, have reflected very different knowledge-constitutive interests. One of these has been that of the academic human geographer – an interest primarily in description and explanation. The other has been that of the urban and regional planner – a contrasting interest in prediction and control. The rationale for these choices and other more general questions as to pertinence will be raised once the approaches have been discussed in their own terms.

Map pattern analysis

Natural and man-made resources and the human activities which utilize and manipulate them always occupy discrete and measurable physical locations. Much of the analytical effort of geographers has for many years now been invested in devising techniques for describing, quantifying and comparing these complex patterns, and well-respected textbooks both at the introductory (Smith 1975) and more advanced (King 1969, Haggett *et al.* 1977) levels are now readily available. It is not the purpose of this section to attempt some sort of cursory review of this work, since it would be impossible to do justice either to its breadth or to its intellectual depth. More important, from the point of view of a general commentary upon applied human science, is to examine the theoretical distinction, if one exists, between the two methodological domains of pattern and interaction analysis. This is because of the paradoxical bias clearly evident in the planning literature but not mirrored in the literature of human geography. Whereas one would expect at least equal significance to be accorded to techniques for pattern and interaction analysis – since traditional planning is concerned just as much with general questions about the arrangements of land uses as with specific issues associated with trip-making – the emphasis has in fact been almost exclusively upon the phenomenon of interaction. Spatial analysis in planning has amounted to very little more than the continuous development and refinement of techniques for describing and predicting different forms of trip-making behaviour.

The reasons for this bias are important and worthy of careful consideration because they are symptomatic of some very basic confusions in planning analysis – confusions between statistical and structural approaches such as were initially discussed in the preceding chapter. In order to highlight and remove these confusions as they manifest themselves in planning applications of the techniques of spatial analysis, it is necessary first to look closely at this pattern seeking aspect of spatial analysis which has attracted so little interest on the part of planners and policy-makers. There must be some inconsistency between the instrumental purposes of map pattern analysis and the perceived spatial information requirements of planners which is simply obscured by a superficial consideration of each, for certainly at a superficial level the consonance between techniques and requirements seems almost self-evident.

Map pattern analysis: assumptions and method

Map pattern analysis attempts to achieve two things: parsimonious aggregation of spatial data and a usually comparative delineation of a discernible shape in their aggregate distribution. Thus the first stage involves an effort simply to describe in a summary and clustered form a subset of the discrete events and structures which are uniquely situated in a given area. The map is the device most commonly used for this purpose, and at least at this level a map-based approach is commonly adopted by planners as well as by spatial analysts. Analysts however do not conclude their investigations when some arbitrary map is drawn. They realize that the aggregation problem and that of pattern recognition are not independent: that maps and other devices for summarizing two-dimensional information are normally produced for the express purpose of revealing or illustrating a pattern. Thus the sorts of questions to which they address themselves typically include the following: what sort of pattern is it that is sought? What sort of aggregation technique will be the most effective for revealing and describing this pattern? And how can the pattern once revealed, and perhaps even quantified, be best compared with other patterns typical of this and other areas?

Of course every map pattern problem is unique, but over the years a variety of generally quantitative techniques have been developed to cope with the simplest and most frequently recurring problems. Multivariate statistical procedures such as factor, discriminant and canonical analysis have proved very useful in the early stages of map pattern analysis. More specific techniques like quadrant analysis, trend surface estimation, spectral analysis and the measurement of spatial entropy and auto-correlation have been employed to highlight certain simple attributes of spatial pattern. These include randomness, clustering and various sorts of auto-regressive data structure.

Map pattern analysis: limitations

Ever increasing statistical sophistication in this area of research has not, however, been an unmitigated advantage. The problem, as noted by Gould (1980, 1981) and Gatrell (1981), is that all statistical and quasi-statistical techniques interpose the computation of functional relationships between the original spatial database (a map for instance), and the simplified description of pattern. Thus the forms of the functions themselves must become (usually unstated) components of the definition of pattern and their particular attributes are clearly hard to justify on theoretical grounds. This has prompted both Gould and Gatrell, and many others besides, to explore the potential of Q-analysis, a technique developed by Atkins (1974) building upon the principles of analytical logic and mathematical topology, for the purpose of isolating patterns which reflect the structure of the interset relationships in databases. One of the advantages of the approach is that measures of pattern or structure are elicited without transforming the original data. One of the disadvantages, however, is that the only form of data acceptable within Q-analysis is that which describes well-defined set-membership relations. In other words, no matter how the information is originally encoded it must be rendered down to a binary scale before it can be accommodated in Q-analysis algorithms. This limitation is only one of several (see Cullen 1984) which might explain the reluctance of planners and other policy-makers to welcome it enthusiastically into their operational tool kits.

Map pattern analysis: applications

The resistance towards more traditional statistical approaches to spatial analysis is harder, on the face of it, to explain. And yet there are, with the one or two exceptions mentioned below, comparatively few examples of the use of such techniques which emanate from a directly practical interest.

There seem to be at least three distinct reasons for this state of affairs. As indicated above, the two main purposes to which map pattern analysis has been addressed are, first, aggregation, and second, pattern recognition and comparison. The aggregation problem in geography is essentially that of zoning or regionalization. How can a heterogeneous set of points or events in space be clustered into a set of zones so that, for example, within zone variance is minimized and between zone variance is maximized? The problem is specified in a variety of different ways but such a formulation is fairly typical. The question is absolutely fundamental to the investigations of the geographer but in planning it rarely arises. Boundaries are in most cases taken as given by the planner. Their establishment is a political event which is comparatively rare. Whereas the geographer can treat the issue of boundary alignment endogenously, as a substantive area for investigation and as contingent upon a series of other substantive research decisions, the planner is almost invariably confined to an acceptance of current boundaries as exogenously defined. When planning boundaries are reconsidered, the issue is almost always an overtly political one. Even the traditional land use plan may eventually become a statutory document, and so the demarcation disputes will be resolved ultimately through political negotiation rather than scientific investigation. This is not to say that such investigation cannot influence the course of any particular dispute. In practice, however, it is rarely seen to be even attempted, let alone influential.

The only exceptions to the rule occur when planning analysts believe themselves justified in removing the spatial aggregation problem from the political arena. This occurs when the political choices appear several stages removed from the question of zoning or regionalization. Thus in a transportation study, the overtly political choices concern the ways in which management and investment policies should respond to likely variations in traffic load upon the multitude of mode-specific links in the transport system. Estimates have traditionally been made by first allocating spatially specific trip generating activities and structures to zones; then predicting each zone's trip generating capacity; then distributing generated trips to destinations; then allocating origin-destination flows to modes of travel; and finally assigning mode-specific journeys to links in the transport network. This and similarly laborious sequences have now more or less faded into the history of the transport planning process, but it at least serves to indicate one of the few circumstances in which direct investigation of the spatial aggregation problem has been seen as of relevance in planning. The relevance is almost always instrumental to what is regarded as the crucial spatial analysis problem in planning, namely that of spatial interaction. Thus the small amount of work that has appeared in the literature (for example Openshaw 1978, Masser and Brown 1977) is almost exclusively dedicated to the design of zoning systems for subsequent inclusion in spatial interaction models.

The second main reason for the lack of enthusiasm amongst planners for research into spatial pattern concerns the traditionally high level of devolution of planning responsibility, at least to the extent that that responsibility involves spatially specific concerns. Thus, in the UK, detailed land use zoning and control is still the responsibility of relatively small local authorities. Apart from the aggregation problem, the other main function of map pattern analysis as already noted is that of recognizing, describing, quantifying and comparing the spatial profiles that may be of interest and importance. If, when planners' responsibilities necessitate involvement with the detail of spatial configuration, they also find those responsibilities increasingly focused upon very small areas, the need for techniques which search for and compare spatial patterns is bound to be limited. Either the areas of their responsibility will be very homogeneous, and thus exhibit little in the way of spatial pattern, or their heterogeneity will be intelligible through intimate personal involvement. The use of sophisticated aggregate pattern analysis techniques would be like employing a steam hammer to crack a nut.

Of course, as in the case of the aggregation problem, there are exceptions. Probably the single largest class are the factorial ecologies which represents a sort of inductive synthesis of the classical ecological models of Park, Burgess and McKenzie (1925), Hoyt (1939) and Harris and Ullman (1945), and the social area analysis of Shevky and Bell (1955). The most common approach is to apply some sort of taxonomic procedure such as factor or cluster analysis to a spatially disaggregated data set composed of a wide variety of socio-economic, demographic and physical indicators (first attempted by Moser and Scott 1961 but repeated, reapplied and extended many times since then). With luck profiles emerge in the form of summarizing variables whose meaning in terms of linear combinations of subsets of the original variables can be interpreted by the analyst. Finally, one or more of these new summarizing variables can be plotted spatially to see whether concentrations of the characteristics which they represent do occur. Such concentrations, if they exist, then constitute patterns which may, of course, be important from the policy-makers' point of view. Recent work in this area (see for example the collection by Herbert and Smith 1979, Holtermann 1975 and Webber 1978) has stressed the policy implications by emphasizing the distributional questions which may underlie revealed socio-spatial patterns. In the UK at least a response by the public sector to such work is evidenced by the shift away from the rather blunt instrument of regional policy, which is by definition insensitive to local spatial variations in welfare, towards much more precisely aimed assaults upon social problems (see Department of the Environment, 1977). Whatever the merits and demerits of such a shift the interesting point from the perspective of this discussion is that it is a shift on the part of central government planners. The geographical breadth of their responsibilities means that overall spatial equity becomes a legitimate concern. They therefore become dependent upon the sorts of map pattern analysis discussed above in order to reduce the spatial and socio-economic complexity of a finely disaggregated nation state economy. The vast majority of those who formulate and implement public policy are, however, working at a local level and so feel no need for such analytical instruments.

There remains a third reason for the lack of interest in map pattern analysis amongst urban and regional policy-makers, and this is possibly the most important of all. It is certainly fundamental in that it relates directly back to the ultimate purposes to which spatial analysis is applied. Geographers as social scientists are concerned essentially with the problems of explanation. They may well accept that any particular solution that they find is bound to be partial, rooted in a framework of cultural, ethical and methodological presuppositions, and thus in no sense neutral nor independent of its context or its subject. They may also employ a battery of mathematical and statistical techniques which lend themselves to a predictive mode of verification. And yet essentially most pieces of work in academic geography stand or fall upon the quality of their explanations.

In planning this is, superficially at least, not the case at all. As was indicated at the outset of the second chapter, planning is about making social choices now so as to ensure a beneficial impact upon the quality of life in the future. It is, in other words, concerned with bringing expectations to bear upon action. Historically sensitive explanations are of course to be treated very seriously. Pertinent and succinct descriptions are of great value. But the *sine qua non* of modern planning analysis is conditional prediction. What the planner can employ most easily is a forecasting device through which the time-dependent implications of alternative assumptions as to the policy and other circumstances of socio-physical collectivities can be plotted inexpensively. In a situation of total uncertainty about the future, any action which has more than immediate effects is bound to appear gratuitous. Yet in the broad field of spatial analysis the vast majority of techniques and investigations which share in common a concern for aggregation and pattern recognition also share in common a disdain for prediction. They fit happily into the descriptive and explanatory moulds of academic social science, but much less happily into the action and future-orientated ethos of public policy-making and implementation. In 1971 Curry expressed his amazement at the complete absence of techniques for extrapolating map pattern descriptions. Since that time some advances have been made (see for example Bennett 1975, 1976, Johnson and Lyon 1973, Martin and Oeppen 1975, Tobler 1973 and Hordijk and Nijkamp 1977), but they have not led to the development of operational models for map pattern projection. Moreover still the main thrust, even of these pieces of work, is towards better descriptions and explanations. The only aspect of spatial analysis in which a major effort has been directed towards the development of predictive devices has been, and still is, that of spatial interaction. It is to this body of research and practice to which we now turn our attention.

Interaction analysis

The analytical treatment of human spatial interaction really became a serious academic preoccupation with the realization that the Newtonian gravity model might, by analogy, also describe aggregate human as well as physical phenomena (see Zipf 1949). This simple device, which described interaction as a product of the two masses and an inverse function of their separation, was obviously a prime candidate for social scientific poaching. Not only did it describe a physical phenomenon – gravitational attraction – which had an apparently perfect social counterpart in the trips generated between spatially separated locations of human activity, but it was also able to achieve such a description without making more than minimal demands upon the data gathering energies of the analyst. The analogy can be quantified using no more than basic census information and an accurate map. The masses become places measured in units of human activity, such as the sizes of their residential populations or work forces, and the inverse distance-squared term becomes a less precisely specified, but usually negative exponential function of the costs of moving within the system.

The interest of planners in this model is no less easy to explain, but derives from more than its analogical convenience and the sparsity of its data requirements. From the planners' point of view such a device is useful as a simple instrument for the generation of aggregate descriptions, but absolutely invaluable if it is also capable of producing reliable projections. More precisely, the planners' policy designing and informing role must sooner or later become spatially specific. Every structure that is built or converted from one use to another ends up as the instrument of a particular human activity only after a more or less lengthy negotiation with planners. Uncertainty about likely shifts in the spatial distribution of activities amounts, therefore, to an inability to participate in those negotiations in a well-informed fashion. It amounts to the nonsense of providing infrastructure and other social capital without the vaguest notion of how it will be used, if at all.

The need, therefore, is for a technique which enables the analyst to treat simultaneously and predictively both human activity and its spatial containment. At an intuitive level this need is obvious. Man-made space is only intelligible in terms of the way it relates to the activities it facilitates. At an analytical level, however, the need for spatially predictive devices to be built upon some sort of humanly functional formalism is less clear. After all the previous section of this chapter demonstrated the feasibility of ignoring, or at least playing down, this link when describing spatial form. All manner of map pattern analysis techniques have been devised which treat space itself, rather than its human uses, as the critical independent variable. We are still concerned here with aggregate analysis techniques, and so there seems no necessary reason why it should be the moments of everyday life rather than the points in three-dimensional space which are the objects of aggregation. In fact the perceived necessity derives ultimately from the requirement that the resulting aggregate device be predictively reliable.

The point is simply that a traditional positivist approach denies the validity of any theory which accords ultimate causal efficacy to aggregate social phenomena. The problem is that any such theory must, at the very least, beg the crucial question as to how these phenomena themselves originated, whether they be cities, classes or religious sects. Thus if we wish to predict the future configuration of an urban area it will be pointlessly circular to treat its current spatial pattern – itself an aggregate social phenomenon of a sort – as the key independent variable. Yet we need to end up with a predictively stable model, and so to invoke the whims and fancies of the people of the area as the key autonomous forces will be equally inappropriate. The way spatial analysts have overcome this problem is through the application of Popper's (1961) doctrine of methodological individualism. According to this argument, aggregate social phenomena become merely the unintended consequences of multitudes of individual actions. Institutions and structures have no existence apart from that of the individuals who comprise them, but these individuals are not themselves consciously and continuously willing the persistence of these social entities. They must, of course, be behaving in mechanically predictable ways which fortuitously ensure the persistence and stable development of macro phenomena, but the relationship necessarily remains fortuitous so long as aggregate deterministic models are thought viable.

Interaction analysis: assumptions and method

The interaction model, as a device for predicting the development of the aggregate social phenomena which have traditionally been of most interest to urban planners, fits very neatly into this methodological framework. Its ostensible purpose is the projective analysis of the spatial distribution of land uses. Since there is no mechanism by which land uses actually manipulate one another the problem must be converted into a micro behavioural one. Thus land uses are reinterpreted as the physical manifestations of purpose-specific categories of human behaviour. It then becomes possible to talk in terms of the activities in which people participate rather than the facilities in which they are contained, and the needs and demands which these activities fulfil rather than the patterns of spatial relationships between different sorts of facility. The simplest interaction models, having adopted this activity-demand terminology, typically involve three fairly basic behavioural premises. The literature rarely mentions, let alone discusses, these premises since it is the ostensible rationale mentioned above which dominates both the detailed configuration of the models and the justification which they receive. However, implicit though they normally remain, these assumptions are critical and deserve mention. They are:

(i)   that people satisfy their needs and demands for the pursuance of activities and the acquisition of commodities by making purpose-related trips to specific facilities;

(ii)   that they exhibit a pattern of priorities as to the order in which these demands are satisfied which is stable, consistent and independent of spatial exigencies;

(iii) that they incur significant opportunity costs in making these trips which are cardinally (or at least ordinally) measurable across population groups.

From these three assumptions it becomes possible to derive most of the simple spatial interaction models which have been, and still are, so popular amongst urban planners. The procedures of this derivation need not detain us long here since they have been very extensively treated in the literature (see Carrothers 1956, Olsson 1965 and, of course, Wilson 1970a and b). Briefly, the most significant of the advances of recent years has been Wilson's replacement of the widely vilified derivation by analogy with Newtonian mechanics with one which rests upon the concept of entropy as employed in the science of information theory.

The paradox of this effort is that, in one sense at least, it may be interpreted as an effort to detach the spatial interaction model from its behavioural roots. Whilst apparently focusing upon the social process of interaction, rather than the simple topography of land uses, the modern entropy-maximizing approach assiduously avoids any explicit reference to individual behaviour. Instead it focuses upon the aggregate constraints which are readily discernible – the number of job opportunities at this place, the amount of disposable income at that. The estimating equation, derived from information expressed at such an aggregate level, is then reckoned capable of describing the aggregate social and physical outcomes of a myriad of individually inscrutable human decisions, without the need for an explicit examination of the assumptions upon which those decisions were based.

In fact it is clear that the premises listed above are just as fundamental to an entropy model as to any other which describes human spatial interaction as a function of the demand satisfying potential of points in space. When we assume that there is an aggregate constraint upon the spatial distribution of retail expenditure in an area – which is defined by the distribution of purchasing power among residential neighbourhoods – we are, in effect, making assumptions about the activity priorities of the residents of the area. Residential choices have all been made and so can form the bases from which distributions of retail expenditure will emerge. Similarly when we postulate a pressing upper bound upon the availability of time and money for retail trip-making we assume that non-negligible opportunity costs are incurred – and are experienced as such – by those who need to make these trips.

Interaction analysis: limitations

Such assumptions may seem unexceptionable. However, recently authors have begun to ask awkward questions about the implications of some of the behavioural underpinnings of this class of model. Curry (1972) first pointed to the dangers of attempting a simple analytic separation between spatial choice and activity choice. Any model, such as that expressed in gravity and entropy equations, which assumes an independent and quantifiable distance deterrence effect, is clearly jeopardized if it is incapable of actually providing a measurement of this effect. Curry has shown that the simple gravity model cannot produce an unambiguous measure of the required sort. The interdependence of map pattern and distance decay effects within the structure of the model means that the distance parameter, when empirically calibrated, will reflect spatial auto-correlation between origins and destinations, as well as any shared response to the physical barriers to communication. It therefore makes no sense to attempt some purely behavioural interpretation of the calibrated term – as for instance a measure of the elasticity of demand for the facilities needed for travelling to a particular sort of destination – since such an interpretation ignores the crudeness of the structure of the model. Yet without any possibility of a behavioural interpretation the model is no more than an arbitrary statistical expression of limited anecdotal interest.

The problem, as Sayer (1976, 1979) has pointed out, is that the model is an historical snapshot of a continuously evolving process of social and economic development. At any one point in time it is a patent nonsense to try and drive a theoretical wedge between a family's slowly evolved circumstantial commitment to jobs, schools and a multitude of other long-term patterns of activity, and the particular event of one member's journey to work on an arbitrarily selected day of the week. The latter is merely an expression of the former, and at least some part of the long-term process would have to be modelled if the daily routine of commuting were to be subjected to a dynamically rigorous analysis. How precisely one would set about this task, and whether or not one would wish to follow Sayer's example and replace the gravity model's simple spatial determinism with an equally simple neo-Marxist economic determinism, are questions to which we shall return later.

Interaction analysis: applications

For the moment it remains to consider whether anything can be salvaged from the interaction modelling tradition. It is, after all, a richly varied tradition, if not at the level of insight then at least at the level of application. The simple gravity model or one of its derivatives still forms the core of most transportation analysis packages. It has been used again and again to describe and predict the spatial distribution of employment, residences, retailing, recreation and many other services and facilities. It has been applied extensively in the analysis of migration. And it normally constitutes the central spatial component of the Lowry model (see Lowry 1964 and Goldner 1971) which still remains as near as planning and human geography have come to producing a comprehensive urban model. Moreover in most of these domains the technique reaches close to the limit of its capabilities without making excessive demands of the analyst for enormously rich data sets.

It would be quite impossible, in a text of this sort, to do justice either to the amount or to the range of effort which has been invested in applying gravity-type models to the practical problems of planning and urban policy formation. There now exist texts offering simplified and exemplified introductions to the practicalities of the model (see for example Foot 1981); texts which review its many uses in practice (see for example Perraton and Baxter 1974); and texts which explore its operational limits (see for example Batty 1976). Its popularity has not, until comparatively recently, been seriously questioned. It has been employed by both the public and private sectors in studies which cover a spectrum extending from metropolitan transportation analyses at one extreme to detailed retail investment feasibility exercises at the other.

Epistemological problems

There are, however, reasons other than lack of popularity for which we might wish to question the use of a given technique of analysis. Ultimately any approach must stand or fall on the basis of its operational efficiency within its own theoretical limits. These limits have the effect of attaching the technique to some specific context in which it may be applied without violating its assumptive base. The problem with the spatial interaction model is that it uses the language of one context – that of social process – to make statements about another, that of spatial form. The physical structure of an area is expressed in terms of human activities such as working, shopping and recreation. Trip-making – another, but purely instrumental, human activity – is derived as a function of the spatial separation of the major activities. Derived trip patterns are then re-expressed as a spatial distribution of the physical containment of each of these activities.

This ambivalence with respect to contexts becomes dangerous when we start to capitalize upon the operational efficiency, simplicity and apparent breadth of scope of the approach. Once the very modest data requirements have been met, the first thing the analyst discovers is that it is comparatively easy to fit the model descriptively to a known spatial distribution of destination activities. The point is that, since a definitional matching between activities and their locations is built into the model, it is quite impossible (as Curry 1972 has shown) to disentangle a description of the social activity of trip-making from one of the spatial auto-correlation of land uses. Thus it is never entirely clear what one is describing when one fits a gravity or entropy model to an observed spatial system.

This lack of clarity becomes positively dangerous when the technique is applied for the purpose which is so close to the heart of the urban planner and to which it appears ideally suited, namely that of conditional prediction. For its mode of expression, in the language of social process, and its theoretical allegiance to the doctrine of methodological individualism are both seen for what they are, aspects of a facade which has the effect of concealing the mechanism of the model. If it is accepted that the model is purely a static representation which confuses the human activity of trip-making with the spatial patterns which both produce and are produced by this activity, then there is only one way in which it can operate as an efficient predictive device. To any extent that the model captures – in a purely statistical sense – the inertia which is literally built into the spatial arrangement of modern urban settings then to this extent it may be capable of extrapolating this inertia into the short- and middle-term futures. In other words, if, despite translating from the language of physical structure to that of social process, the model nevertheless achieves an efficient description of the underlying structure which itself exhibits the required inertial stability, then it may be capable of providing certain limited sorts of relatively reliable forecasts.

The limit upon the scope of these forecasts is, of course, a fairly tight one. The success of a prediction when it is achieved is an indication that very broadly defined spatio-temporal trends have continued unabated and despite the deliberate attempts of planners and others to divert them. For this reason the only sort of future uncertainty that the planner may reasonably hope to reduce by using spatial interaction models is that which is contained in such trends. He or she may perhaps ask of the gravity model: what is likely to happen if I do nothing or if I am unsuccessful? What will happen if I do this, intervene deliberately in that way or introduce the other policy are all, however, quite illegitimate questions, because each denies the premise upon which the success of the technique depends – namely the unfettered continuation of past trends. It is only when a behavioural interpretation of the parameters of a model is feasible that it becomes possible to use it for conditional projections or an analysis of the impact of policy. And as we have seen such interpretations are not permissible in the case of gravity and entropy equations. A model built upon a theory of the behavioural dynamics of a particular activity setting might conceivably be used for the assessment of the effect of policy since it would at least be couched in appropriate terms. It would thus be amenable to questions about the way in which individuals and families might react to changes in their circumstances occasioned by the activities of the public sector. As we have seen, the gravity model depends critically upon a purely statistical description of aggregate spatial pattern and thus has nothing to contribute to an analysis of the processes whereby policy actually effects changes in that pattern. In epistemological terms, therefore, it is essentially indistinguishable from the map pattern techniques discussed earlier in the chapter.

Synthesis

As just one of many techniques for the statistical analysis of spatial form, it none the less remains easily fitted and simply operated. Moreover, it provides a reasonable descriptive summary of aggregate spatial patterns and permits straightforward extrapolative projections. All this is achieved with an estimating equation which is characterized by fairly gross specification error and is thus seriously liable to misuse. How should the spatial analyst respond? There appear to be at least three alternative, though not necessarily mutually exclusive, strategies worthy of consideration. The first and simplest would be to throw out the spatial interaction model altogether as a dangerously misleading device whose successes may be achieved less pretentiously by the use of more straightforward map pattern techniques. Another strategy would be to develop the method further, using entropy and other approaches, so that a dynamic and disaggregated series of spatial interaction models might result which avoided the specification errors of the simple models and were based upon firmer behavioural and structural premises. A final alternative would be to retain gravity and entropy models as deliberately crude aggregate and statistical devices purely because they provide a simple way of translating spatial form into social process and vice versa, so permitting the vertical integration of spatial and aspatial analysis in planning. Before concluding this chapter we shall consider each of these strategies in a little more detail. Space is still the central medium of planning control, and therefore analysis which is spatially specific would appear to be of continued relevance. However, it is by no means clear that this analysis should take the form of an extension of either of the approaches discussed in the two main sections of this chapter. We have indicated why planners typically demonstrate a far stronger preference for interaction analysis techniques than for those designed to describe map pattern, but we have also shown that this preference is founded upon a basic misconception. The choice between alternatives for the future should therefore be based upon a serious reassessment of the general questions raised at the outset of this chapter. We should no longer ask simply which technique works best, since this is a meaningless question, nor indeed which responds most readily to the practical requirements of the planner, since this is the sort of question which has in the past provoked the almost fanatical enthusiasm for gravity and entropy models. We must rather ask whether indeed there exists any real need to perform aggregate spatial analysis, and, if such a need can be demonstrated, what it means to perform this analysis in the different ways alluded to above.

This question is an open one only because it refers specifically to aggregate analysis. The planner or policy-maker is involved, as an agent of the state, with some negotiated concept of social welfare. Any such concept, no matter what its ideological roots, derives its relevance purely from the way it relates to the shared lifestyles and aspirations of one or more client populations, which cannot themselves be rendered effective and intelligible without the consumption of space. Patently, therefore, there must be a spatial dimension to the understanding upon which a planner must build his or her policies – and this must be true even for those policies whose ostensible concern is not with explicitly locational issues. However, the crucial question for this discussion is whether or not any sort of aggregate analysis of spatial forms and distributions is capable of contributing to a relevant understanding of the impact of space upon the lifestyles and needs of urban and rural communities.

The question can be resolved only by means of a clarification of what we mean by space. To any extent that space – or more broadly the overall physical form of an area – can be appropriately regarded as an absolute, to that extent it should be susceptible of aggregate analysis, since to that extent it represents a structure whose impact is independent of social meanings and human interpretations. And of course in the limit space – or at least a fairly simple space-time relativity – achieves a sort of absolute structural status. In the short term – given a certain technology and a constant urban form – there is a limit to the distance a person can travel in a unit of time. There is a limit, which is the same for all individuals, to the number of routes which may be traversed, the number of destinations which may be approached, and the number of activities which those destinations are capable of supporting. We can, for instance, usefully measure the capacity of an airport in terms of departures and arrivals in a given period of time. We can specify the range of effective choice open to groups of individuals with respect to facilities such as schools, shopping centres, libraries and health centres. We can estimate accessibility levels – in terms of distance, time, money or some combination of these units – of one place to each of a series of others. Moreover, as Hagerstrand and his colleagues have demonstrated (see for instance Hagerstrand 1970 for the first statement of the case), it is possible to devise a generalized language and set of techniques for achieving time-space analyses of these sorts. The point is, however, that this language and these techniques bear very little resemblance to the spatial analysis procedures discussed above. They do not purport, as do many of the interaction models, to aggregate from individualistic theories of spatial choice. Nor do they approach spatial behaviour, as do the map pattern techniques and – in practice – the interaction models, as a problem of simple statistical description. In so far as they remain aggregate techniques, they do not treat behaviour directly at all, but rather the limits to behaviour. In other words, they concentrate upon the careful and comprehensive analysis of the setting and the context (see Lenntorp 1976) rather than the ways in which they are used.

The problem is, of course, that space is only absolute in a crude, though sometimes importantly limiting, sense. Thus the most illuminating of spatial analyses accept that though form is related to behaviour, this is only after the mediation of the meanings that are attached to it through the perceptions of individuals and groups. In other words, space becomes effectively relative to its human interpretation, so that its impact upon aggregate behaviour must be treated as essentially problematic. No longer is it sensible to search for simple mechanical relationships between any absolute measure of spatial form and the behaviour it contains, for the relationship is not of that type. We must now accept that, whilst spatial distributions of human activity may be describable through aggregate analysis of the gravity and entropy variety, they may be interpreted only via some form of human analysis at the micro level, for it is at this level that the determining relativity of space is injected.

Some of the many implications of this simple proposition will be addressed at a later, more appropriate, stage. For the moment we are concerned purely with the question of what, if anything, should be the role of aggregate spatial analysis, and how, if at all, it should be performed. It seems that, apart from a particular type of spatio-temporal structuralism which sits unnaturally at any point on the aggregate-disaggregate spectrum, the only safe application is in the area of simple description. Certainly the whole range of approaches discussed in this chapter, from the less popular map pattern techniques to the widely used interaction models, are all subject to this limitation. They all work by summarizing something of the aggregate topographical form of spatial behaviour rather than the processes which are responsible for that form, and so can at best be employed predictively by relying upon the inertia of the topography and extrapolating a relatively accurate description of it.

Given this limitation, the first of the three alternatives listed above seems an obvious choice. If the description of spatial patterns is all we are currently capable of achieving, given the range of available techniques, then what arguments can be adduced for selecting techniques which have potentially misleading pretensions in other areas? If when we appear to be describing the way people choose to distribute their purpose-related trips what we are in effect doing is using the language of social behaviour to describe the historically determined patterns of urban forms, then why should we want to venture beyond approaches which address such patterns directly? The argument, in other words, for selecting techniques for map pattern or spatial auto-correlation analysis rather than gravity or entropy models is simply that they are theoretically clear. They describe what they purport to describe. Moreover, in that interaction models are not in their simplest form, capable of describing much else, it is clear that even for predictive purposes the map pattern techniques are likely to be just as successful.

The possible counter arguments are of two sorts, and each suggests one of the two remaining alternatives proposed at the start of this section. It may be argued, first, that the critique outlined above is not characterized by objections in principle to spatial interaction models, but only objections to them in their crudest forms. Thus, possibly through the further use of entropy-maximizing procedures, it may be that these objections can be overcome. Theoretically explicit disaggregations and sophistications of other sorts can be introduced into the basic model to add the element of behavioural realism which is lacking in its crude aggregate form. Thus Curry (1972) has suggested a way of separating a so-called 'pure gravity' effect on trip-making from the autocorrelation between origins and destinations, employing a feedback loop in the equation system. And others (see for example Wilson 1970b and 1972 for early examples which are still much discussed) have proposed a variety of specific disaggregations for precisely specified uses of the model. In each case the sophistication is designed to give voice to a theoretical statement about the behaviour or circumstances of the individuals constituting the aggregate whose patterns of movement are the subject of the investigation.

At one level, each of these elaborations should be assessed upon its own merits and defects. Some will represent more apposite theoretical propositions than others. Specific consideration of some of the propositions will be made at later points in the book. However, at this stage the key question concerns the technical validity of the process of embellishment. Though it is of course dangerous to generalize on this point, since the variety of the sophistications which have been proposed for the simple gravity model is considerable, it is probably safe enough to assume that all must face one basic problem. This is simply that as each novel refinement gently coaxes the gravity model away from its aggregate bluntness it cannot avoid simultaneously detaching it from the source of most of the successes it has achieved. If the model is, for theoretically important reasons, limited to providing a certain sort of broad statistical description of an unscramblable mixture of the reciprocal impacts of form on behaviour, then disaggregation will normally be counter productive since it will amount to a behavioural elaboration of a theoretically ambiguous core equation. Whatever that equation represents, it is certainly not a straightforward theory of human behaviour. Behavioural disaggregation of gravity models may be based upon the most interesting and illuminating empirical research or upon the most sympathetic and original theoretical speculation. It may produce elaborate equations which calibrate better than their aggregated equivalents. The problems of interpretation are still not reduced.

A very simple example should make the point clear. The most basic singly constrained gravity model is one which, using entropy procedures, is normally derived from two sets of constraints, one defining the trip-generating potential of a set of origins and the other describing some aggregate limitation upon the capacity of trip-makers to make longer and longer trips. The estimating equation which results describes the probability of interaction as directly proportional to the distance to each destination. The probability of a destination attracting a trip is thus said to be independent of its size. The aggregate statisticians will regard this as an inefficiency in the estimating equation and no doubt try a variety of destination size or capacity indicators until a better fit is achieved. The disaggregators, on the other hand, will theorize, no doubt very wisely, that trips of different sorts are attracted to particular destinations partly as a function of the attractions which those destinations have to offer. Thus the way in which individuals respond to distance in selecting a destination must be theorized explicitly and this theoretical sophistication must then be built into the estimating equation. The point of quoting this hypothetical example is simply to show how inevitable it is, when attempting a disaggregation of the basic model, to drop back into the parlance of social process. Even though the model is not one of social behaviour and even though the parameters of its crudest form are quite incapable of any legitimate behavioural interpretation it becomes necessary to assume such an interpretation before those parameters can be elaborated, embellished and disaggregated. And though the theorizing which motivates and informs the elaboration may be valid, the gravity model is simply an inappropriate place to start.

If this second alternative is therefore dismissed, there remains finally the possibility of accepting the mis-specification of the crude models, accepting their descriptive limitations and yet retaining them, at least on some occasions, in favour of the less confusing map pattern techniques simply because they do describe spatial form in terms of the human activities it helps to shape and reinforce. The arguments for such a strategy all derive ultimately from an admittedly pragmatic view of the role of aggregate spatial analysis. If it is assumed that the urban planner's interest in urban form and pattern is always instrumental in achieving some ultimately social purpose which has to do with the human activities that may be facilitated or impeded by particular physical configurations, then the force of this third alternative starts to become apparent. For if it is accepted that at least at the aggregate level it is quite impossible to force an analytical break in the circle which links form and process, then most simple aggregate spatial analysis is bound to remain statistical in the sense described above. Certainly the two large classes reviewed here achieve no more than alternative static descriptions of the topography of form and behaviour. Neither represents a testable theory of the behavioural dynamics of urban form. Theoretically, therefore, there seems little to choose between the two, and so the next logical step appears to be that of asking operational and pragmatic questions. Which approach fits best with the available evidence? Which provides the most useful descriptive output? And which links most readily into the overall framework of urban planning analysis?

The question of fit is of course one which cannot be resolved here and since there are *a priori* arguments on both sides there probably is not much in it one way or the other. The argument for map pattern approaches is simply that much of the descriptive success of the gravity model is due to the map pattern which underwrites the behaviour measured, and so it is much less confusing and much more straightforward to quantify that pattern directly. There is then no danger of calibrating the model to an arbitrary and transitory set of behavioural coincidences which are of no long-term interest whatsoever. The counter argument is, of course, that since one can imagine a spatial system, characterized by absolutely no auto-correlation amongst origins and destinations, in which a gravity model could still be usefully calibrated, there are bound to be some occasions on which map pattern techniques are of little or no use whilst gravity models fit well.

Since the mind recoils in horror at the thought of a precise empirical test of the relative goodness of fit, the chances are that the task would not be worth the effort. A clearer answer, however, appears possible to the questions of practical utility. At the level of simple description, the output measures derived from the use of each approach are very different. On the one hand, a map pattern analysis will, no doubt with great efficiency and accuracy, give the analyst a clear indication of the extent of spatial auto-correlation of a given zoning system. The gravity model, on the other hand, will describe an actual pattern of social interaction which, though no doubt an indissolubly partial product of the map pattern, is of immediate relevance to the information needs of the planner in the way that an index of auto-correlation will never be. From a trip distribution matrix a whole series of accessibility indices may be computed – indices which lend themselves to political comparison, indices which are immediately intelligible to the layman, and indices which are expressed in the language planners are wont to employ.

Nor does it stop at the level of simple description. Planning as aggregate analysis comprises several distinct stages which fall fairly naturally into a logical sequence. Thus it makes no sense to talk about the spatial distribution of employment growth (or decline) until the overall structure of the local economy and its aggregate growth potential have been reviewed. Similarly spatial disaggregation of residential demand presupposes analysis of the likely future of the demographic base. The requirement that a programme of aggregate analysis should be hierarchically integrated in the manner suggested by these particular rules of thumb has as its most obvious corollary that the language used in each step should pose the minimum of translation problems when it comes to the passage between steps. The clear advantage of the gravity model is of course that in (however dubiously) employing the language of social process it fits readily into a framework defined at a higher level by socio-economic analyses of transaction and population structure.

By far the best-known analysis procedure which capitalizes upon this facility is of course the Lowry model, which is, in effect, two gravity equations hierarchically linked through the use of a crude economic base mechanism. However, this is only one of many ways in which spatial specificity may be introduced to a programme of economic and/or demographic analysis and forecasting. The point of interest to this discussion and at this point is simply that, whatever the nature of the modes of analysis chosen for the aspatial phase of the linked process, a gravity or entropy model is almost certain to be preferred over some form of map pattern analysis for the spatially specific phase purely because it describes and extrapolates the spatial allocation of economic and other social activities rather than the fabric which accommodates them. The distinction may seem a fine one, but those who have attempted any sort of translation between the languages of form and process will probably acknowledge that it is in fact profound. The straight logistics of the task would be enough to put off all but the most courageous of planning analysts. The point is that the forecasts produced by devices such as the input-output and cohort survival models can, with comparatively little manipulation, be treated as the exogenous triggers for the extrapolative use of gravity and entropy models in a way which is impossible if straightforward map pattern techniques are employed.

Figure 4.1 is an indication of just one of the methods of achieving vertical integration within a programme of aggregate planning analysis. It extends and lends spatial specificity to the programme outlined briefly at the end of the preceding chapter. As in that case it is neither intended as a comprehensive model nor as exclusive of alternative approaches. It is rather a partial approach to the problem of mounting and cross-referencing two simple and parallel sequences of macro statistical description and extrapolation. The feedback mechanism (such as the responses of unemployment and/or migration rates to aggregate imbalance between demographic and employment forecasts) and autonomous filters (such as the various public sector responses to spatial allocation forecasts) are not in any sense 'built into' the integrated package. The statistical nature of the projection techniques precludes any such simple strategies for closing the system. These boxes in the diagram must in effect be understood as representing analysis procedures which operate at a different level. Any outputs derived from mechanical forecasting devices such as those hooked together in the diagram will inevitably mean nothing – either for analytical or practical purposes – until they are interpreted by the analyst or policy-maker from their respective standpoints. It is essentially this process of interpretation which is referred to by the broken line boxes in figure 4.1.

What the diagram clearly illustrates is the relatively simple way in which parallel aggregate analysis of the economic and demographic bases may, via the use of interaction models, be rendered spatially specific. The input-output technique enables a simple three-way reaggregation of employment classes into those whose locations are least likely to be related to local patterns of demand ('export employment'), those whose locations are most likely to be correlated with sites of other, functionally linked, economic activities ('intermediate employment') and those whose locations are most likely to be correlated with the spatial distribution of residences ('induced or retail demand'). By the same token, aggregate demographic forecasts can, by means of measured activity and household formation rates, be reaggregated into workforce and household estimates. Assuming that a recent census is available, the existing spatial distributions of these employment and residential activities can be established fairly accurately. It simply remains to calibrate and run a set of five simple interaction models, which, as we have seen, interrelate these sets of activities by describing trip distributions, in order to derive spatially specific forecasts for each of the major categories of aggregate social activity. Moreover, the other relevant demand forecasts – such as those for various recreational activities – can probably be produced by simple extensions to the basic statistical framework.

It is worth noting that this flexibility is what lends the approach outlined above a potential which is not shared by the traditional Lowry model. First, the use of an input-output formulation of the local economy, rather than the crude economic base mechanism, permits analysis which is more sensitive to the idiosyncrasies of its productive structure, is richer at the level of description, and is thus likely to be more

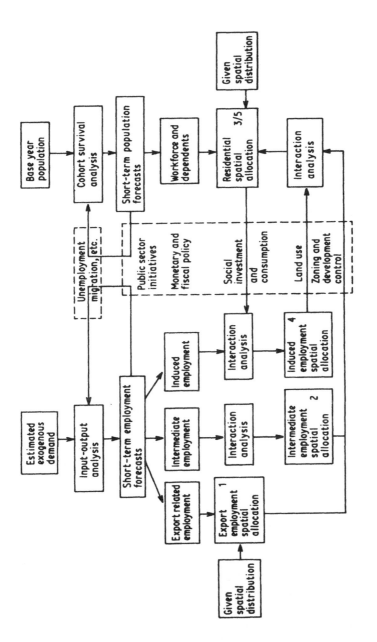

4.1   Spatially specific integrated aggregate analysis

accurate at the level of extrapolation. Second, the explicit inclusion of a cohort model to represent the local demographic base – but without formally linking it with the input-output model – forces the analyst to consider the nature of the inter-dependence between the economic and population bases. In other words, as indicated above, the analyst is required to stand deliberately and explicitly outside the statistical method of forward extrapolation in order to employ the results of any forecasts. Nothing is concealed, as it is in the Lowry model, through the use of arbitrarily mechanical and theoretically equivocal negative feedback loops to render inconsistencies consistent. And finally, the fine grained functional classifications characteristic of input-output and cohort models permit the calibration of a more realistic set of spatial interaction equations than is possible with the Lowry model.

It may, in the not too distant future even, prove possible to formalize something of this process of analytical integration. Further formalization will not, however, be of any use if it is interpreted as the definition of a set of rigid mechanistic links between existing aggregate models. One way of thinking of the problem of integration is to start by defining the limits of such a process of mechanization in the light of the critical discussions of the previous sections and chapters. If we accept that the main techniques discussed in this and the preceding chapter – input-output, cohort survival and interaction analysis – together represent the limits of our current ability to formalize the study of aggregated economic, demographic and spatial demand structures in a purely mechanistic fashion, then it follows that further formalization – particularly of the problem of integration – must take some non-mechanistic form.

One possible approach currently attracting a great deal of interest in the spheres of medicine, engineering and geology, is to adopt some of the methods developed in the fields of artificial intelligence and knowledge engineering. Many advances have recently been made in this area of applied computer science (see Feigenbaum and Barr 1982 for a recent and committed review). For the purposes of aggregate analysis in planning and human geography, however, perhaps the most significant development has been that of the expert systems model. This is effectively no more than a means of applying an inference system, generally based upon the principles of predicate calculus, frame theory or semantic networks, to a data or knowledge base for the purposes of problem solving, diagnosis or hypothesis testing. I have elsewhere (see Cullen 1983) discussed the early stages of an experimental project designed to make use of such a model in planning and architectural education. However, its power to act as the core of a sort of loosely structured envelope within which the aggregate techniques might be applied as an integrated and internally consistent set, appears considerable. The chief advantages of the approach seem to be:

(i)   that the data or knowledge base can accommodate a far broader range of types of information, including high levels of uncertainty, intuitive connections, patchy local knowledge and the results of partial sub-models, than is permissible within traditional mechanistic frameworks;

(ii) that this knowledge base is strictly independent of the inference system, so that the latter does not need to be modified every time the structure of the former grows or changes, as is the case with most traditional approaches;

(iii) that the use of the system is transparent rather than opaque to the user, so that as a programme of analyses and results are being assembled, the user can ask for an explanation which will take the form of a demonstration of the way in which any given point in the analysis process is related to the current goal (which might for instance be the estimation of future net migration levels), hypothesis or problem diagnosis.

It would, however, be foolish to get carried away with the richness of the aspatial and spatial analyses that are, or may in the future be, permitted through the use of a loosely integrated procedure of this sort. As noted in the preceding chapter whatever the beauty of the pictures painted by input-output or cohort survival models, they make no statements whatsoever as to the dynamics of economic and social behaviour. For predictive purposes the structures described are thus the only independent variables required. Patterns of behaviour are assumed either constant, or varying more or less randomly within limits which ensure the perpetuation of the initially described structures.

It is exactly the same in the case of the interaction models discussed in this chapter. Whatever the ostensible differences from and practical advantages over map pattern techniques, since no theoretical statement is made about the dynamics of spatial choice, each approach amounts to the same thing, statistically, when used for prediction. Each effectively relies upon just one set of independent variables, and that is the set which describes directly or indirectly – the land use and network topographies of the area in question. Again it does not actually matter what individuals choose to do so long as they either choose to do exactly as they have done in the past or vary their choices in a way which is consistent with aggregate constancy and thus the perpetuation of the existing topography. Thus for predictive purposes, no matter which of these spatial analysis approaches is chosen, our capability is limited to that of extrapolating from existing structural patterns. Of course, as with input-output and cohort survival models, it is important to attempt some sort of measure of our confidence with respect to any given forecast through the use of Bayesian and other techniques. Moreover, semi-formalized integration procedures, perhaps based upon some of the principles of the artificial intelligence models discussed above may in the future offer more relevant and realistic predictions. But this should not be allowed to deflect our understanding of the ultimate source of the uncertainty which such strategies are designed to circumscribe. It is to an understanding of the processes which generate this aggregate uncertainty that we must now turn our attention.

5

Perceptions, values and lifestyles

In this chapter, we shall, in effect, go back to the beginning once again and examine the analysis problem afresh. The premise upon which the initial discussion was based will, however, now be relaxed. To this point the fundamental assumption has been that a programme of planning analysis could be devised in a coherent and practical fashion without making explicit theoretical statements about the processes of urban and regional dynamics. In this and the following chapter we shall treat the mechanisms of urban change directly, and, in so doing, develop and elaborate the argument that such a task requires shifting one's focus from the aggregate to the disaggregate level of scrutiny. Thus, we shall argue that in order to understand the processes of urban dynamics one must examine directly the practices of social choice. For it is none other than these practices which mark the discontinuities in the lives of their participants. And it can be none other than such discontinuities which collectively constitute the patterns of urban social and economic change and development.

Up to this point, and by ignoring dynamics, we have revealed an analysis programme which is capable of providing structural descriptions of local and regional economic and demographic bases; statistical approximations of a variety of types of spatial pattern; trend extrapolations of each of these descriptive pictures; and a limited measure of integration amongst such elements of a comprehensive aggregate analysis package. The purpose of this and the next chapter is to initiate a discussion of various methods and theories available to assist a search for the dynamic processes which generate and transform the patterns isolated through such a programme of static analysis. We shall start by treating approaches which completely reverse the logic of the last two chapters by adopting, at least as their ostensible focus of attention, what is, for all practical purposes in planning and human geography, the irreducibly disaggregated subject of study – the individual human being.

More precisely, we shall consider the methods, and their theoretical roots and assumptions, proposed by the human sciences specifically for the study of the perceptions, evaluations and lifestyles of unique individuals acting in roles relevant to the interests of planners and urban policy-makers. The rationale for this dramatic reorientation is obvious. If we are eventually to examine the choices which add up to change in urban patterns and processes then we must first consider the ways in which those choices – ultimately the choices of individuals – arise. Without begging too many basic questions, it seems reasonable to assume that since choices are in the last analysis made by individuals, they must to some extent arise from the ways in which those individuals impose meaning upon the contexts in which their choices must be made. In other words, they arise partly from the ways in which contexts are perceived, evaluated and used. The element of idealism in this position will be compensated to some extent in the next chapter by an examination of methods which acknowledge that choices are not, for the most part, made by isolated and autonomous individuals, but are generally negotiated within and between groups whose socially defined and maintained roles – whether they be family, friendship, class or institutional roles – inform the practices and outcomes of the choices they make.

For the present, however, we shall discuss approaches, not to the issue of choice in all its social complexity, but rather to those of its building blocks which appear to require an unambiguous concentration upon the individual. Though we shall treat each of the substantive concepts of perception, evaluation and use separately, this typology does not reflect a well-formulated theoretical demarcation, but rather a convenient way of classifying the ostensible foci of the most popular methods of investigation. In other words, what follows is essentially a methodological discussion. This is because the research in this area has been characterized by a fairly sharp methodological cutting edge, but one which is supported by the most insubstantial of theoretical handles. The reasons for the theoretical weakness of work in the area all no doubt boil down to the unavoidable truth that the human mind is an inscrutable subject for study. But this does not explain the effort that has been invested in the development of sophisticated measurement instruments, especially amongst social scientists such as geographers and even economists, working at the academic end of the planning and urban or regional policy domains. For such individuals a micro level enthusiasm is not necessarily one of their terms of reference.

There do, however, appear to be at least two very good explanations, both for the interest in micro level research amongst geographers and planners in the late 1960s and early 1970s, and for the methodological lines along which that interest was pursued. For the purposes of the present discussion it will be necessary to review them but briefly. The first – which I have discussed at greater length elsewhere (Cullen 1976) – relates to the academic pedigree of the interest. The 'quantitative revolution' in planning and geography of the early 1960s soon began to creak at its theoretical seams, and one early response was an attempt to inject 'behavioural realism' into the crude aggregate models which constituted its first operational creatures. Thus since the need was for modifications and embellishments to quantified tools of analysis, the response had to be provided in an appropriate language. And so, whether theoretically respectable or not, sophisticatedly quantified measurements of attitude and behaviour became the order of the day.

Perhaps even more important, however, was the political as opposed to the more narrowly academic roots of a blossoming interest in micro level research. For at about the same time that geographers were becoming frustrated by the aggregate crudity of their mathematical models, so planners and others actively involved in urban policy formation were beginning to question their political ones. In particular, they were beginning to question the complacent neutrality of the simple rational model of decision-taking, in which normative statements are expected to emerge more or less self-evidently from a mixture of abstract and anodyne goal statements and a positivist analysis of the *status quo*. Again it would be inappropriate to enter a lengthy discussion of the issues involved here, partly because I have treated some of them already in chapter 2 of this book and elsewhere (see Cullen 1978); partly because there is already an extensive literature on the subject anyway (see, for example, Rein 1976 and Scott and Roweis 1977); but mostly because the important point for our current purposes is simply to note the major outcome of that period of academic, political and professional soul searching in the late 1960s and early 1970s. That was the participation movement (reviewed by Fagence 1977 and Thornley 1977). One obvious way of coping with the distributive implications of planning decisions and their resultant political nature, given the inadequacy of checks and balances built into the formal model of representative democracy, is, so the argument goes, through direct public involvement in the process of decision. Its outcome may be sensitized to public opinion, and the political responsibility thereby devolved, by means of direct public participation.

Opponents of the movement would no doubt use the word 'confused' rather than 'devolved' when talking of its effect upon power and responsibility, and it is certainly true that no one in authority has even seriously suggested democratizing local policy formation in the tradition of Plato's *Republic*. Public participation has generally been discussed and implemented as at best a sensitizing instrument of paternalism. The chief stumbling blocks have thus been logistic and economic. How may we cheaply, speedily and accurately tap opinion upon this or that topic? Once again, therefore, interest in micro level investigation has been stimulated. Once again methodological sophistication has been the paramount requirement. And once again the purpose has been that of informing planning and policy formation, but this time directly at the point at which the decision is taken rather than indirectly through the embellishment of its analytical base of macro statistical models.

We now turn to a discussion of the products of this two-fold enthusiasm employing the theoretically weak but methodologically useful typology referred to above.

Perceptions

Studies of the ways in which people perceive the world in which they live have a history which is, in one sense at least, longer than that of social science itself. They have been pursued in an enormous variety of ways, from the totally informal to the fully quantified, and have been predicated upon an extremely broad range of epistemological positions. Despite this apparent flexibility, their popularity in human geography in the 1960s and 1970s was never then – nor has it been ever since – matched by anything approaching an equivalent interest in planning and other policy-related fields of study. This is probably because they possess no obviously direct political relevance. No matter how they are performed, they remain essentially descriptive of cognitive structures. They do not, therefore, lend themselves to straightforward co-option by policy analysts. Of all the approaches that will be reviewed in this chapter, perception studies appear to have the least to offer the politician in the sense alluded to above. However, political convenience is not the best test of a branch of applied human science, and there is clearly a sense in which the study of environmental cognition is critical to an understanding of urban processes and problems.

Perception analysis: assumptions and method

At their simplest, perception studies are designed to reveal the way in which an individual or group makes practical sense of some aspect of the external world. The theoretical questions raised by such an endeavour are often ignored in the presentation of the results of perception studies, but most of the better-known studies in human geography rely – whether explicitly or implicitly – upon a model of cognition in which concepts such as 'mental map' or 'perceptual schema' play an important role. The characteristics of these pre-conscious devices vary with the predilections of their commentators, but some of their more popular features are commonly held to be the following. First and foremost, they are almost invariably thought of as possessing an organizing or structuring function. In other words, they are said to act as frameworks through which individuals classify and render intelligible the barrage of sense data received from the external world. In particular this ordering function is designed to enable us to navigate paths through space; to locate ourselves and other subjects and objects relative to each other; to coordinate our activities in space; and sometimes (though this is stretching the generalization) to facilitate the formulation of normative and comparative judgements of aspects of the physical and natural environments. It is worth noting, moreover, that interest in the mental map is often focused not upon its success as an organizing instrument, but upon non-random indications of its failure. In other words it is sometimes cited as an argument in favour of the existence of the mental map that its power of systematic distortion may be empirically demonstrated. The best documented examples are on the topic of distance estimation for which mistakes have been systematically related to direction (Lee 1970) and more generally to familiarity (Stea 1969).

On the shape rather than the function of mental maps, agreement is harder to find, presumably because the theoretical questions raised, and left without satisfactory answers, are more precise. However, many studies have adopted some variant of Kelly's personal construct theory (see Bannister and Mair 1968 for a full discussion) and thus have assumed them to have some sort of bipolar configuration. In other words sense data are thought to be classified and interpreted through the use of a set of contrasting attribute pairs. The park is near, or it is far away. The wall is high or it is low. The walk is long or it is short. Moreover, many authors allow mental maps to be composed of normative as well as purely descriptive attribute pairs. The park is friendly or it is unfriendly. The wall is attractive or it is ugly. The walk is pleasant or it is unpleasant.

Perception analysis: limitations

The literature on perception studies contains many extensive discussions of the shape and functions of mental maps (see Downs and Stea 1977 for one of the most comprehensive reviews of work in the area). Moreover, the specific issue of their bipolar configuration has proved of particular interest, but more as a methodological rather than theoretical question. The literature thus suggests that the central problems or limitations of perception studies are methodological ones, and these, broadly speaking, fall into one of two categories. How may we define the stimulus most clearly, and how may we measure the response most accurately.

The different answers which have been offered to the first of these questions vary greatly with respect to their technical ingenuity, but without noticeably reducing the impact of the basic logistic problem of confounding effects. The aim is generally to approximate an experimental design in which the stimulus is held constant for each respondent. In the complex world of urban environments this means relying either upon a verbal account of the stimulus of interest or upon the use of some sort of non-verbal surrogate. In the first instance the analyst must face up to all of the many problems associated with the relativity and ambiguity of natural language. The very use of such language involves generalization and that in its turn implies imprecision as to the denotations of words and phrases. The use of surrogates like photos, films or even three-dimensional models to overcome these difficulties is in fact simply to replace the confounding effects of one non-neutral medium of communication with those of other media whose properties as artefacts are equally incapable of dissolution or neutralization.

Even if it were possible to ensure that the stimulus was held constant for each respondent, the same class of problems would still have to be faced in attempting to measure each response. And if we wish to make precise inter-personal comparisons amongst those responses, the problems become even more intractable. What appears to be the easiest way of making such comparisons – namely via quantification – has certainly been the target of extensive methodological investigation. Analysts have employed rating scales, rank ordering procedures, semantic differential and repertory grid techniques all in an effort to gather precise, generalizable and comparable measurements of aspects of people's mental maps. And yet we are still not much closer to answering the question of what it means to ask an individual to convert both affective and cognitive components of a multi-dimensional perceptual model into points upon a fixed length cardinal or ordinal scale. The alternative of relying upon graphic measurement techniques (usually by asking respondents to draw upon base maps) has been far less popular, largely because it severely limits the range of questions one can in practice ask of the mental map concept. However, the medium is no less subject to the problems of interpretation alluded to above.

The independent and confounding effects of the medium are of course at their most acute when the analyst attempts to measure the affective components of the mental map or schema. The problem is not the unfamiliarity of the medium for in the case of the map's most subjective dimensions there is no choice but to use the languages of value, emotion and volition, and they (unlike, say, the Ordnance Survey base map) are fundamental to the practice of everyday discourse. We all share a workable commonsense understanding of what is meant when a house is described as attractive, a shopping centre as convenient and a park as fun. It is not that the words are themselves unfamiliar but rather that the rules governing their application are much more contentious, much more difficult to specify.

Part of the problem is certainly to do with the hierarchical nature of language and the patterns of perception it reflects. It is generally accepted that the affective levels of perception are, at least in part, a function of the simpler cognitive levels. We change our feelings towards persons or objects as our beliefs are adjusted to reflect what we have most recently learned. Thus if we are asked to select from a list of ten cities that one which we regard as the most attractive it would be illegitimate to assume variation in responses to be a simple product of varying affective dispositions. It may equally be due to varying knowledge of the cities in question. Perhaps much more important, however, is the problem of interpreting variation in affective response even in the absence of cognitive ambiguity. The point is that for the most part it appears grossly unrealistic to regard an attitude of any sort as unrelated to specific human purposes. A city may be attractive to visit but less so to work or live in. Yet the typical perception study is not normally set within a theoretical framework which actually links attitude measurement with purposive behaviour. The relationship is either ignored or assumed as unproblematic. Rarely if ever is an attempt made to measure it.

This is perhaps the crucial weakness of most perception studies. It is not that their central theoretical device – the mental map or schema – is totally inappropriate, for whatever its nature (about which we still know remarkably little) there must be some sort of framework through which our experience is filtered and by which it is organized. The real difficulty is that any such concept is doomed to failure when used as the sole and independent theoretical device within a self-contained empirical study. Probably the most crucial requirement of any micro level research is that it should investigate patterns of attitude and behaviour which make transparent sense to the subjects of that study themselves. Thus it is absolutely essential to relate both perceptions and evaluations of an environment to the ways in which that environment is used. Research treating the delineation of a mental map as its end product can all too easily deteriorate into little more than a study of verbal behaviour or map-drawing ability, detached from all the areas of investigation and understanding that are capable of lending it pertinence. Far too much of the research in this area seems to rest upon an unstated assumption that the mental map may be regarded as a sort of neural equivalent of a computer – an information-processing device capable of investigation without regard to the interests of the person using it.

Perception analysis: applications

It is probably not for this reason, however, that mental map studies have not been employed extensively in practice. We have argued above that there emerged during the 1960s in the form of the participation movement a clear role for some means of quantifying attitudes to the environment. The movement provoked a multitude of empirical studies (see Fagence 1977), the vast majority of which took the form of questionnaire surveys. However, almost without exception, these exercises shunned the format and approach adopted by so many in the – particularly geographical – research community. Instead they mostly took the form of more or less elaborate hypothetical questions. At a superficial level, this seemed to be an appropriately pragmatic response to the problem of tapping grassroots political opinion. It wasted neither the time of the respondent nor the effort of the policy analyst upon the intermediate, and thus marginally relevant, issue of environmental perception. Instead it addressed the crucial issue of comparative evaluation directly.

Values

Of course as we have seen, many perception studies, in treating the processes whereby people assimilate and organize sense data, have included amongst those processes some which are explicitly affective or normative as well as the more narrowly descriptive ones. In this sense, therefore, such perception studies have also become studies of evaluation. However, it was noted further that some of the serious technical and theoretical problems of perception studies were at their most acute when they attempted to address affective components of the mental map or schema. More generally, to any extent that such studies do treat the assimilation of sense data without further considering the ways in which that assimilated information is employed in the choices people make and the lifestyles they adopt, to that extent the normative and evaluative dimensions of the studies must be the most suspect ones. For whilst it may, on occasions, be permissible to regard a description of an object as partially independent of the purpose to which that object is to be put, the same can surely never be said of evaluation. Our feelings towards a particular park or shopping centre must be very closely related to the ways in which we have used those places in the past and what we want from them in the future.

One particularly glaring manifestation of this difficulty of treating perception studies also as studies of evaluation arises when one acknowledges the relativity of most evaluations. Because evaluations are so intimately related to the choices we make, one of their most important attributes is that they treat their objects as interdependent. In the complicated world of everyday life a choice will almost invariably involve some sort of trade-off between competing alternatives. Accordingly, the methods and language we have evolved for the purposes of evaluation recognize this fact and treat the issue of appraisal as a relative rather than absolute one. By contrast, and presumably because of their failure to consider the determining relationships between perception and action, most mental map studies treat each evaluation as autonomous and absolute, and their objects as unrelated. They have thus remained ultimately uninteresting from the point of view of the planner or policy-maker.

Value analysis: assumptions and method

There are techniques, based generally upon hypothetical questions or simulation, which attempt to overcome these problems by attacking the issue of evaluation in a radically different way. Rather than treating the unobservable and affective dimensions of behaviour directly, they attempt to derive them through a carefully controlled and interpreted study of observable behaviour. In this way a major defect of perception studies – namely that they risk irrelevance by separating the treatment of attitude from that of action – is apparently side-stepped. The gaming studies, by working back from actions – or choices – to the attitudes upon which they are based, can never be accused of ignoring the relationship. And what this means in practice is that the importance of trade-offs, of relativity amongst evaluations and of interdependence amongst their objects can all be duly acknowledged in the implementation of the techniques.

The techniques as they are now employed in studies such as those of Hoinville (1971), Rowley and Tipple (1974) Rowley and Wilson (1975) and Jones (1979) amount to a fusion of work in two distinct traditions – that of laboratory-based research in experimental psychology coupled with that of revealed preference analysis in micro economics. The purpose of the fusion has almost invariably been that of directly informing the process of policy formation rather than that of adopting the less direct route of first of all improving its analytic base. For this reason the techniques are designed primarily for the purpose of deriving quantitative measures of aggregate social preferences. More specifically each application involves something like the following three stages of research design:

(i)   define a set of related objects – some of which are presumably social goods – for which it would be pertinent to establish some sort of cardinal social preference function;
(ii) define an individual or family level choice which will both make sense to those asked to participate in the exercise and involve as a by-product a trade-off evaluation of at least some of the objects defined in stage (i);
(iii) design a one-person game which involves participation in a series of hypothetical choices of the sort defined in stage (ii) in such a way that the trade-offs between some or all of the stage (i) objects are expressed in a cardinal currency.

If subsequently samples are well designed and the fieldwork well organized then the problem of computing aggregate social preference functions becomes an apparently trivial one. In practice the generality of the above outline has, however, been matched neither by a correspondingly broad range of applications nor by the clarity and pertinence of the aggregate preference functions derived.

Value analysis: limitations

The explanations are several. First, there are good reasons for believing that the responses elicited in a gaming exercise of the sort described above may not accurately reflect the evaluations of the individuals consulted. The hypothetical question is in general one of the most difficult both to ask and to answer, and all gaming studies are built ultimately upon a series of hypothetical questions. More specifically, the nature of the technique is such that there appears to be a serious danger of eliciting either trivialized or systematically biased responses. It may be, for instance, that participants will adopt standardized responses to situations which they define as games. These responses may be prompted perhaps by the desire to win, to speculate, to fantasize, to be reckless, etc. It is generally reckoned that these tendencies are most likely to emerge when the mechanisms of the game are complex and thus distract the respondents' attention from the basic issue. If, however, the game is pared down to an irreducibly minimal form of hypothetical question, then it ceases to be a game and so detaches the issue from its everyday behavioural context. In such circumstances the shock effect of a blunt hypothetical question – how much compensation would you require for the loss of your home? – may well have an equally serious distorting effect upon responses.

There is generally no such shock effect, of course, if the potential magnitude of the impact of a planning decision is not spelt out. Probably the most common form of preference question employed in applied research is of this latter variety. In other words, it leaves the respondent to draw out the connections between his or her pattern of life and the implementation of a proposed policy. It simply demands an evaluation of that policy. It loses all aspects of a game, but in so doing also drops any pretence to the claim of behavioural realism. The point of the gaming formulation is to translate a decision which has to be made at the level and in the terms of the urban policy domain down to the level and the terms of everyday life. Thus there is no distorting shock effect because the issues raised in the hypothetical question are not seen to have any immediate or significant personal relevance. And in such circumstances, responses are likely to be somewhat arbitrary.

Whether or not hypothetical questions are put in the form of a game, responses may further be distorted in a systematic fashion if participants perceive – whether correctly or not – a connection between their participation and their own self-interest. Thus if a game or a question is inexplicit about the costs of the public goods under discussion – parks, policing, etc. – then respondents may overestimate their preferences because they judge that they will never have to pay for such goods themselves. If on the other hand, the schedule is both explicit and realistic about such costs (and perhaps includes questions about willingness to pay) then respondents may underestimate their preferences, because in this case they judge that as individuals they cannot significantly affect the outcome of the study but there is always a chance – whatever the protestations of the interviewer – that their replies may eventually be used to determine the way levies are charged.

Even if one has good reason to believe that the responses received have been neither trivialized through the mechanisms of the game or the perceived irrelevance of the questions, nor biased systematically as a result of a strategic approach by subsets of the participants, there are still serious difficulties to be faced in attempting an interpretation. The aim of the planner or policy-maker in using such a technique is normally – as noted above – that of deriving one or more aggregate social preference functions for a variety of social goods or externalities. The problems of interpreting aggregated responses to gaming studies in this way are due partly to the difficulties implicit in the necessity of making interpersonal comparisons and partly to the nature of the functional forms computable.

By and large it is not too serious an oversimplification to say that the difficulties of making interpersonal comparisons increase monotonically with the power of the unit of measurement adopted. Thus a series of simple binary choices may be aggregated into a social preference function without much difficulty. Ordinal scaling presents serious problems of interpretation (as demonstrated in the work of Arrow 1951 discussed briefly in chapter 2). And cardinal scaling, whilst perhaps overcoming or obviating some of the problems of an ordinal scale, by introducing a currency which is independent of the issue of choice, introduces also all the difficulties associated with that currency itself. Some gaming studies employ only ordinal scales. However, the majority select some form of cardinal currency – usually money or time – in order to lend realism to the exercise. Thus a respondent may be asked to choose a location for a house by selecting a set of neighbourhood attributes each of which is priced as perhaps an increase in rent or in capital cost (see Hoinville 1971). Alternatively a family might be asked to speculate upon how they would vary their everyday lifestyles – measured in cardinal units of time per activity – if some change in their pattern of accessibility were to occur (see Jones 1979).

The trouble is that the currency chosen will itself be accorded a value by each respondent. And this independent, and from the perspective of the study, irrelevant valuation may vary systematically across the sample and thus distort systematically any aggregated result. It is easy to see how this may happen when a monetary unit is chosen since individuals will possess a command over the currency which varies with their real wealth, and we should reasonably expect their marginal valuations thus to fall as their wealth increases – giving rise to what is commonly known as the phenomenon of the 'rich man's pound'. When time is chosen as the unit of account, the position is more complicated since it is only over the long term that command over this currency appears to vary systematically – in this case with age rather than wealth. However, even over a twenty-four-hour cycle there are many reasons for suspecting the neutrality of time. Two very obvious examples of such reasons are, first, that its availability, and thus perhaps its value, varies from day to day and person to person with the nature and level of their long-term commitments. Thus a working wife with young children may find an extra ten minutes of travelling time far less tolerable than would a single person approaching retirement. Secondly, though each of us clearly does 'possess' no more than twenty-four hours a day, increasing wealth can increase the productivity of that time (by for instance providing faster modes of travel, time-saving gadgetry and the means whereby the less satisfying commitments of life may be reduced or eliminated through the ability to buy services). We shall turn in the next section to more basic questions concerning the explicit treatment of time. It is sufficient here simply to note that the use of any cardinal currency in a simulation study introduces more difficulties than it resolves.

The general difficulty its use is supposed to resolve is that of translating between individual and social choices. This becomes possible, first, because the currency is capable of straightforward aggregation and, second, because, if that currency happens to be a monetary one, the aggregations of a series of individual valuations are directly comparable with the discounted costs of the various alternative public investment or consumption policies. Thus social choice becomes a fairly mechanical exercise of benefit-cost appraisal (see Lichfield *et al.* 1975). We have seen that aggregation itself raises many problems of interpretation. However, even if these may be overcome (perhaps by means of a weighting procedure or by restricting studies to homogeneous samples) the subsequent problems of comparative evaluation are by no means trivial ones. A full treatment of the problems of public sector investment appraisal techniques is beyond the scope of this work (see Merrett and Sykes 1973 for a comprehensive discussion). Here attention is focused upon evaluation problems that derive directly from the means whereby information is collected. A crucial general problem derives from the indivisibility of the public goods which normally constitute the focus of gaming or other hypothetical choice studies. Such social products as amenity improvements, rescheduled bus services or newly provided public open spaces are not typically paid for directly by their consumers. Moreover even in the case of private goods and services the consumer is almost invariably a price-taker. Thus any realistic simulation study is bound to employ a design in which each benefit is quoted to respondents at a fixed price. The difficulty then is that an aggregation of the responses gives one no clue as to the shape of the demand curve for the facility or service in question. All we learn about is one point upon that curve, and though that point may suggest an aggregate willingness to pay which is less than the discounted cost of provision, this finding is purely relative to the arbitrary quoted price. If the price had been either raised or lowered, it might, depending upon the price elasticity of demand, have suggested the opposite conclusion. The only way around the difficulty is to increase dramatically the sample size and quote different prices to equivalent subsets within it. This will generally be an unrealistically expensive option.

Value analysis: applications

Despite the limitations, the hypothetical question and its various sophistications have long been popular in planning and policy studies for the reasons noted earlier. They seemed to represent a way of short-circuiting the route between policy formation and decision, avoiding at least some of the need for conditional forecasting. Given the political interest in participation exercises as sensitizing devices, many concluded that at least some aspects of policy impact could be estimated directly – by asking hypothetical questions of the likely victims – rather than indirectly by means of formal models designed to estimate social responses in a purely mechanical fashion. And so questionnaires employed in planning have on many occasions fallen into this category of hypothetical or simulation studies. They have shown little interest in generalized perceptions or current lifestyles, but have instead focused directly upon questions of the form: 'what would you do (or think) if ...?' As we have seen this is perhaps the most difficult sort of question in the survey research tool kit, both to ask pertinently and to answer without risk of misinterpretation.

Of course the problems which the planners face are not hard to understand. Current perceptions and lifestyles are not necessarily good indicators of the ways in which people will respond to non-marginal changes in their future circumstances. Yet it is for the evaluation of just this sort of proposed change that planners need information, and something like the hypothetical question appears to be the only simple means of providing that information.

In fact, as with perception studies, the information generated by a simulation or hypothetical choice study, however conveniently adapted to a simplified model of the policy-forming and informing processes, is capable of serious misinterpretation due to the fact that it reflects and describes just one link in a complex cognitive-behavioural chain. Mental map studies fail when they ignore the patterns of behaviour which are constructed as it were to give voice to the assumptions and perceptions which constitute the cognitive devices in question. By the same token hypothetical choice studies fail when they ignore the motives, assumptions and meanings which underwrite the specific activity sets which respondents are asked to reconstruct hypothetically. The failure in each case results through failure to realize the essential circularity of the relationship between the observable and unobservable levels of human behaviour. The choices and lifestyles in which we participate are as inseparably the products of our attitudes and perceptions as these attitudes are themselves the products of experiences assembled in the course of living out our choices. To attempt the investigation of abstracted attitudes detaches that study from all that lends those attitudes importance or even relevance. And to attempt the investigation of abstracted behaviour – of which a hypothetical choice must be an archetypical example – is, no matter how convenient from a practical point of view, tantamount to succumbing to exactly the same fallacy.

Lifestyles

Clearly the most straightforward response to the practical and theoretical difficulties attached to gaming studies is to drop their most problematic element, namely the hypothetical question. The traditional revealed preference study adopts this strategy and explores only the choices that have already been made, not those of the hypothetical future. This general approach has been the staple diet of empirical micro economics and market research for many years (since, in fact, the original statement by Samuelson 1947). Studies in these areas have typically concentrated on retail consumption and expenditure profiles. In mainstream sociology analogous studies have concentrated rather upon activity patterns at a cross-national or cross-cultural scale using some form of time-budget analysis technique (see for example Szalai 1972, Michelson 1977, BBC 1975, Robinson 1978 and Chapin 1974). Since in planning and urban policy analysis the interest is generally directed towards the consumption of public goods and services, the sociological focus upon activity patterns has tended to be more popular than the economic one upon expenditure and purchasing. But in either case the general purpose has been strictly analogous to that of the gaming studies discussed above. Techniques have been developed to provide quantified estimates of social preferences through the observation, measurement and aggregation of a series of individual choices.

Revealed preference analysis: assumptions and method

The central problem, noted above, with any study based upon hypothetical questions is that there is no way of ensuring the realism of the exercise in which the respondent is asked to participate. We cannot know, in presenting an individual with a hypothetical dilemma, how the resolution that we are offered fits into the context of that person's everyday life and the framework of assumptions upon which that lifestyle itself depends. Thus researchers who prefer to explore preferences revealed retrospectively through the choices an individual has actually made do so primarily because this appears to be the only way of ensuring the required framework of circumstantial realism. If people are assumed to be rational, and well informed and in a position to make a reasoned choice, then the choices they make must – so the argument goes – be the best evidence of the feelings they espouse.

Unfortunately the conditions attached to this proposition – namely those of rationality, knowledge and unconstrained choice – are central to the utility of the approach. Any analysis of behaviour designed to reveal people's preferences can normally do no more than weight equally all of the options apparently available to the group in question. Thus if an individual participates in one available activity and not another it is assumed that he or she has deliberately chosen the former and knows about the latter. Moreover it is further assumed that that person entertains a stable and consistently ordered ranking of the available options and that his or her knowledge is sufficient to make such a ranking. Without such assumptions the concept of a social preference function becomes a nonsense. When one reflects upon the processes of consumption and activity choice it becomes clear just how unrealistic these assumptions may be. Preferences of the 'right sort' may very frequently simply not exist.

Even if they do exist, however, the technique assumes further that no circumstance, such as a spell of bad weather or a bank overdraft, will intervene to distort overt behaviour away from its chosen path. If in practice such patterns of behaviour are products of their contexts as much as of the process of deliberated choice, problems of interpretation will be very difficult to overcome in the use of the information collected.

Revealed preference analysis: limitations

Clearly, this is the central operational problem for the designer of any revealed preference study. As we have noted, the model of choice generally adopted is one which probably does not correspond, in its emphasis upon complete knowledge and simple-minded rationality, to much of what we understand as the reality of personal choice.

Perhaps even more damaging to the potential of such studies, however, is the possibility that even if appropriately well-informed and rationally organized preferences do exist, overt behaviour will fail to reveal them. The reason is that, in attempting to ensure the realism of a systematic measurement of preferences, these studies inevitably end up by measuring much more than the undistorted products of individuals' preferences. Overt behaviour is in most cases clearly not just a mechanical response to a consistently organized rank ordering. It is also the product of the social, physical and temporal environments in which it is located. The absurdity of assuming otherwise may be perhaps most clearly demonstrated through the example of activity choice (the only sort of revealed preference which is likely to be of interest in the non-market domains of urban and social policy formation). If a group of individuals is asked to record in sequence all the events of a twenty-four-hour period, then typically the majority will describe a day comprised of between twenty and thirty discrete episodes. If subsequently these accounts are subjected to a revealed preference analysis, then this will almost certainly involve treating each of these episodes separately as evidence of the activity preferences people espouse. It will probably also involve treating the frequencies and durations of these episodes as measures of the relative strengths of those preferences. Thus we are presented with an absurd picture of life in which individuals sit back every fifteen or twenty minutes and reflect upon their activity preference functions in order to come to a decision about whether or not to continue with what they are currently doing or switch to something else which has become – for some unfathomable reason – marginally more preferable.

Revealed preference analysis: applications

All this makes it difficult to interpret diary or lifestyle records as indices of revealed preference, but not totally impossible. There is clearly a sense in which the aggregate amount of time dedicated to a purely discretionary activity counts as an index of its popularity. If two groups of people with similar commitments and facing similar patterns of accessibility spend their free time in quite different ways then it would seem not unreasonable to conclude that their preferences differ.

Probably the safest applications of the technique have been those conducted at the broadest of scales. The best examples include cross-cultural comparisons in the Szalai (1972) tradition, and broad longitudinal comparisons like those of Gershuny and Thomas (1981) in the UK and Robinson (1977) in the United States. It is not that they strictly satisfy the conditions noted above, but rather that multitudes of specific infringements cancel each other out in the very large samples and over the very long periods typically covered. Thus, if the theoretical background work has been well done, major differences in activity patterns may be legitimately interpreted as cultural shifts which are not purely context dependent. However, for the reasons noted above, a cautious approach to such interpretations must be the rule. Moreover, what may be legitimate at the broadest cross-cultural or longitudinal scales, may well be quite illegitimate at the urban short-term scale at which the distorting effects of misinformation or intervening circumstances may be systematic. None the less, the painstaking work of Michelson (1977), which will be considered more closely in the next chapter, is evidence that useful results may be achieved at varying scales.

Time geographic analysis: assumptions and method

Perhaps the clearest contrast to the approach of researchers such as Szalai, Chapin, Michelson and Robinson is that offered by the

Lund school of geography, especially during the period of Hagerstrand's greatest influence in the late 1960s and 1970s. Over that period a quite different interpretative framework was provided for the study of everyday lifestyles, labelled by its authors 'time geography'. This mixture of definitions and descriptive graphic devices laid by far its heaviest emphasis not upon the preferences and choices which are reflected in everyday lifestyles, but upon the objective circumstances which serve to channel, circumscribe and, in the limit, precisely fix the patterns which they exhibit. In his original statement of the model, Hagerstrand (1970) suggested a three-way classification of these constraints: first, capability constraints which range from biological necessity through to physical inaccessibility; second, coupling constraints which define where, when and for how long the individual has to join other individuals, tools or materials in order to produce, consume or transact; and, third, authority constraints including national frontiers, private property rights and bank opening hours. The impact of these constraints is typically documented through the use of a two-dimensional time-space map. This plots in vertical sequence (see figure 5.1) each of the fixed activities in a person's day (fixed in one of the three senses noted above) and thus indicates any flexibility through the use of essentially residual time-space

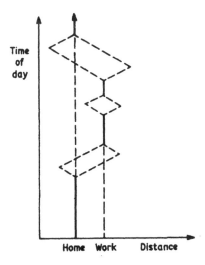

5.1   The prisms of an individual's twenty-four-hour activity programme

'prisms' which define the feasible 'region' of activity between two fixed events. Over the years since Hagerstrand's original regional science paper, the school has grown in size and strength and the language of time geography has increased in scope and sophistication (for a recent review see Parkes and Thrift 1980), way beyond the point at which we could hope to do justice to its rich variety in a review of this sort.

This process of expansion and diversification clearly bears witness to the potential of the language and framework. It also underlines the fact that it is ultimately no more than a set of tools for the description of an activity or lifestyle in a way which focuses our attention upon the centrality of the concept of spatio-temporal continuity. It is therefore a little misleading to talk of specific assumptions and limitations since these vary from study to study as the language is extended, applied and reapplied to new areas and new issues.

Time geographic analysis: limitations

None the less, just like any other language, that of time geography is in no sense a neutral medium. It lays emphasis, as we have noted, upon the role of circumstantial constraints in the determination of human lifestyles. This emphasis has been reflected in most of its applications, and thus constitutes its major limitation. It offers a powerful medium for documenting the impacts of context, whether that be the context implicit in the rules and routines of family life or that contained within the political and economic structure of post-industrial society. It does not offer a way of understanding how the deliberated actions of self-conscious individuals can manipulate the framework itself because it concentrates upon the separation rather than dynamic integration of the two levels. Though the limits to discretion are clearly of enormous practical and political importance, an approach which simply explores these limits adroitly sidesteps at least some of the key theoretical questions.

Time geographic analysis: applications

Its practical relevance has not, however, gone unnoticed. Apart from the many specific studies undertaken within the Lund school itself, the overall influence of the approach has probably been as great as that of any other major methodological initiative in post-war human geography. It would therefore be quite out of the question to attempt to review the field here. It is none the less probably worth mentioning one piece of work that has been employed in many practical contexts (including applied transportation research in several countries), and that is the development of the PESASP model by Lenntorp (1976). This is designed to describe the activity sequences or time budgets that are feasible given the full range of time-space constraints of the sort noted above. The model has been used to investigate policy issues such as access to medical and child care facilities, as well as for more narrowly defined transport questions.

Epistemological problems

The central theoretical problem with time geographic and diary-based lifestyle research, just as with the perception and evaluation studies reviewed earlier in this chapter, really boils down to one of methodological or linguistic myopia. Perception studies were shown to be weak primarily because the methods employed focused all analytical attention upon the psychological construct itself – the mental map or perceptual schema. The crucial relationship which renders the mental map pertinent to the perceiver and capable of reliable interpretation by the analyst – namely the link with activity and choice – is generally ignored. In a similar fashion various simulation techniques have used artificially constructed hypothetical choices in order to reveal people's comparative evaluations of features of their environments. The most important relationship ignored in this case is that between the individual's perceptions, attitudes and personal circumstances on the one hand and the choices which emerge from reflection upon that context on the other. To abstract the choice from its reflexive context is to trivialize it and thus to jeopardize the chance of documenting in a reliable manner the ways in which people impose value and personal significance upon their environments.

Studies of people's lifestyles, whether from a time geographic or revealed preference perspective, are equally inadequate. A focus upon the observed pattern of experience is clearly consistent with a multitude of different and possibly inconsistent theoretical interpretations. To say that that pattern is a reflection of its author's choices is, in at least one sense, true since in the limiting case all constraints are breakable. Even the prisoner in a high security jail is 'free to choose' the alternative and possibly terminal lifestyle that will result from a hunger strike. To argue conversely that in a complex post-industrial metropolitan environment lifestyles are inevitably a reflection of their circumstances is equally unexceptionable, and equally uninteresting. The point is that the diary technique is just a way of assembling a description and time geography just a way of organizing and presenting one. Neither addresses head on the critical theoretical question as to how, in practice, the real conflict between our reflexive capacity of choice and our historical accumulation of commitment is resolved day by day through the determination and experience of a lifestyle.

Synthesis

On the other hand, and despite their failure to resolve what may prove to be fundamental philosophical difficulties, both diary and time geography research highlight the ground over which the battle is likely to be fought.

For the present therefore we shall ignore the subtleties of the philosophical discussion (and return to them in the concluding chapters of the book). Instead we shall explore the possibilities of a more practical integration of the two approaches – possibilities which derive simply from their shared interest in the patterns of day-to-day lifestyles. The feasibility of such a project is suggested by a close reading of the theoretical discussions of influential researchers in each tradition (for instance Chapin 1974 and Hagerstrand 1974). For though their emphasis, and thus the ways in which their methods have been developed and applied, are different in the ways noted above, neither school precludes the approach of the other in a strict theoretical sense.

Reconciliation is thus likely to be most simply achieved by specifying the partiality of the model of human behaviour addressed directly by each group. Rendering both down to their irreducibly simplest forms results in the contrast, which is an inevitably gross caricature, depicted in figure 5.2. In each case behaviour is seen as a mechanical product of just one broadly defined independent variable, either internal and unobservable or external and observable. The simplest way of reconciling the two approaches is clearly by means of direct combination, so that behaviour is treated as some function of two rather than one independent variable. The relationship remains unidirectional and thus completely mechanical.

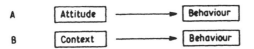

5.2   The revealed preference and time geography models of human behaviour

The difficulty with this solution is that it is based upon a theoretical position which reflects a less clear rather than a clearer understanding of the phenomenon of human behaviour. The resulting model amounts to no more than the proposition that behaviour is not totally without cause or reason. Anything within the universe of observable and unobservable phenomena may constitute part or all of that cause. And on the very reasonable Popperian principle that the only good theory is a parsimonious one, this theory must count as a very bad one indeed.

Moreover, such a simple-minded approach ignores, as does each of the original submodels, the essential circularity of the relationships under scrutiny. On the one hand the simplest revealed preference model, in treating behaviour as an unmediated uni-directional function of attitudes, completely ignores the process whereby attitudes are themselves formed. Just as our motives, assumptions and preferences serve to define and direct the lifestyles in which we participate, so also do those lifestyles fix the domains of experience in which predispositions and attitudes are generated. Usually the easiest way to explain attitude differences is thus in terms of different experiences. No one would expect two people to express a similar attitude to travelling by bus if one had never made such trips in the rush hour and the other had never made any trips other than at such a time.

On the other hand, a naïve time geography model ignores the fact that the contexts which constrain human behaviour are in fact by and large man-made. Thus, though patterns of behaviour may often appear adequately 'explained' by reference simply to the contexts in which they are located, those patterns must, sooner or later, contain actions whose intended or unintended consequences are modifications of the very contexts in which they occur. Thus, whereas at one level the timing and venue of an evening meal may be adequately explained in terms of family structure and residential location, at another the argument that constitutes that meal as a social event may precipitate the end of the marriage, and thus the dissolution of the coupling constraints implicit in the old family structure.

Of course, if one concentrates exclusively upon the events of the immediate present then each of the separate simple models may provide technically adequate explanations (of mode choice in terms of attitude differences and of the timing of meals in terms of coupling constraints). This, however, is not the point. Each of the micro-level approaches reviewed to this point have been shown to be inadequate primarily because of their myopia – perception studies failing because abstracted from a realistic behavioural context and hypothetical choice studies because abstracted from a realistic framework of meaning. Each has failed because it has ignored the circularity of the relationship between attitude and behaviour. The study of everyday lifestyles, perhaps through the use of some form of diary method, appears to be a more appropriate starting point because any discussion of perception or attitude is then automatically attached to the reality of experience. Moreover the focus is upon actual rather than hypothetical behaviour. But as we have seen, a method which directs our attention exclusively towards very brief slices from the lives of a group of individuals is capable of all manner of interpretations.

What is required is an appropriate theoretical framework, one which makes explicit the dual circularity noted above and summarized in figure 5.3. One way of elaborating and integrating these circles is through the explicit treatment of time. Several of the studies performed at University College London over the past few years (see Cullen and Godson 1975, Cullen and Phelps 1975 and Cullen, Hammond and Haimes 1980) have demonstrated that, over a daily or twenty-four-hour cycle, the revealed preference model of activities as the product of deliberated choice is an almost totally inappropriate one. Day-to-day working lifestyles are heavily routinized. Moreover this process of routinization appears to be an almost unavoidable response to the complexity of the contexts of urban life – the arrangement of the physical and temporal environments, the overlapping frameworks of commitments, and the baseline of economic resources – all of which persist over a much longer than daily cycle. In one sense, therefore, the time geography framework appears by far the most acceptable.

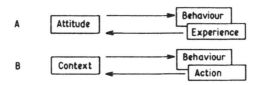

5.3 Reflexive versions of the revealed preference and time geography models of human behaviour

However, significant components of this overall context which defines working lifestyles are self-selected. In other words, the routinized process observable over a typical twenty-four-hour period represents not just an adaptive response to long-term circumstances beyond the control of the individual. It is just as likely to reflect adaptation to circumstances whose configuration has been determined over the years by a sequence of deliberated 'life' choices – about where to live, whether or not to start a family, which job to go after, etc. Each of these choices has direct (or realized) and indirect (or unrealized) consequences with respect to everyday activity patterns, and it is, amongst other things, the difficulty of rendering these lifestyle implications consistent with one another so as to produce an acceptable daily activity pattern that entails the high level of routinization which we have observed.

Thus the paradoxical circularity of the relationships between behaviour, attitude and context, noted above, can only be made intelligible if we elaborate the circles in such a way as to differentiate clearly the long term of persistent structures and destabilizing choice from the short term of routinized behaviour and repetitive experience. Figure 5.4 attempts to do just this. It is not intended to be a fully comprehensive model, either of long-run or of short-run behaviour. The point is simply to demonstrate how each is unintelligible if viewed in isolation from the other. For just as our long-run plans and attitudes define, through the choices we make, the context which fixes the daily routines of our lives, so those routines themselves provide us with the experiences upon which subsequent plans are built and new attitudes formed. To talk of circularity in this way is not, of course, to imply any sort of mechanical closure of the model. Even in the crude form depicted diagrammatically in figure 5.4, the model is constructed so that, from the perspective of the individual, there are at least three general sources of independent influence. First, the physical, social and economic circumstances in which long-term decisions are taken are shown as having an independently constraining effect upon the results of those decisions. Second, the process of adaptation is described as conforming to its own rules and procedures. Thus an individual's pattern of daily routines becomes a mediated function of the context in which they evolve – the mediation taking the form of a sort of self-constructed programme of adaptation. And finally – perhaps most important of all – the plans and interpretations which inform the crucial life choices which an individual makes are seen not as a simple function of a circumscribed pattern of experience but as one to a significant extent controlled by a process of creative self-reflection.

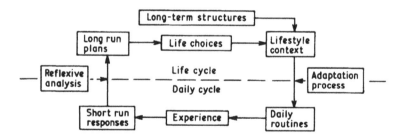

5.4   Reflexive adaptation model of day-to-day lifestyles

If the above is accepted as a reasonable theoretical starting point, it still remains to examine the utility of specific studies of everyday lifestyles – perhaps employing some form of diary technique – within the framework which the theory defines. On the face of it such studies look much more immediately appealing than do perception or simulation studies since they alone focus upon that critical link in the circle at which the physical   and   non-physical   environments   are   actually experienced. By observing and documenting day-to-day lifestyles we are thus recording the evidence of choices made in the past, the events to which immediately present feelings and attitudes are attached, and the springs of choices yet to be made. However, the problems of interpretation are still considerable.

The University College work referred to earlier has lent general empirical support to the theoretical framework proposed above but it has also highlighted the difficulties of generalizing from research pitched at the day-to-day level. First, these studies have repeatedly confirmed that at the time scale of everyday experience the process of routinization is such that it becomes quite impossible to unscramble the long-term relationship between choice and circumstance that is actually reflected in the events of a typical week or day. Thus though we may have good reason to believe daily lifestyles to be adaptive responses to important and carefully deliberated life choices, we have no means of drawing conclusions about these choices from accounts of those lifestyles if indeed we also believe other long-term processes and circumstances to be of determining significance as well as those of choice.

A further corollary of the above theory, and one which is also now well-documented empirically and which presents further problems of interpretation, concerns the subjective levels of day-to-day experience. It would appear that the process of adaptation which entails routinization also entails a sort of defensive narrowing of the attitudinal spectrum, a pre-conscious strategy of self-protective blinkering. The result is that, with certain exceptions, only the more negative responses of stress and distress regularly penetrate the curtain of routine and thus remain available and capable of being recalled for subsequent discussion. There is also some evidence of a displacement effect which can distort even this pattern of negative response. In one study, for instance, variation in work-place stress and distress levels correlated more closely with variation in the length and complexity of the journey to work than with any measured attribute of the job itself, and yet the work journey was rarely itself described as stressful or unpleasant (see Cullen 1978). Of course one may only speculate statistically about such unacknowledged relationships. However for those who might wish to build a broad-ranging and sophisticated environmental attitude investigation on to a diary study of daily lifestyles – on the reasonable grounds that those lifestyles constitute the vehicles of environmental experience – the above findings offer very little encouragement. And so an entirely laudable intention to ground attitude studies more firmly in context will be thwarted if it is considered that the individual's routine patterns of experience define that context.

6

Structured locational decisions

The argument introduced at the outset of the previous chapter remains to be elaborated and discussed. In that chapter it received what may be described as circumstantial support. To reiterate, the central proposition was that the dynamics of urban structure can only be approached analytically by a direct examination of the decision processes of individuals, acting in roles relevant to planning interests. It is these decisions which mark the transition points in the histories of the institutions of urban society and the biographies of their members, and collectively these transitions constitute the processes of urban change. The proposition holds for institutions as different as a simple nuclear family and a multi-national corporation.

The preceding chapter offered what I have called circumstantial support for the argument in the following sense. A discussion of the micro-level building blocks of complex social changes – the perceptions and evaluations from which they stem and the lifestyles which they define – demonstrated that no one of these aspects of human behaviour was itself independently critical in determining the outcomes of complex choices. In other words, no matter what we may learn of the process of choice, of the accumulation of experience, of the shaping of assumption and interpretation, research in any one of these specific areas will always leave us well short of a comprehensive account of the outcome of long-term choices. There are really two reasons for this conclusion, each of which emerges from the discussion of the preceding chapter. First, the circularity of the relationship between long-term choice and short-term experience means that an understanding of one will always beg an understanding of the other. And second, the circle is open, rather than closed, in at least three important respects, so that a study based, say, upon one of these independent sources of influence (perhaps the interpretative process of attitude formation) will always leave the outcome of some dependent event within the circle (long-term choice being the obvious example) uncertain. The other independent influences of context and adaptation process may well have an equally crucial role to play. These influences upon the circular relationship between choice and experience may be legitimately treated as independent only because they are in this instance being viewed from the perspective of the individual living out the consequences of his or her own choices. Clearly, from other perspectives the context of any given style of life is susceptible of manipulation as are the reflexive processes of adaptation and interpretation.

Of course, this is all rather negatively circumstantial considered as support for the above proposition. For it does not amount to convincing proof of the dynamic significance of micro level choices. It is purely a demonstration that no simpler micro level of research (into for instance attitudes or lifestyles) is capable of generalization to the level of urban dynamics. It offers simply a broad indication (no more) that an appropriate investigative starting point may well prove to be that of the complex social choice which seems theoretically pivotal. An obvious immediate response would be that the theoretical model presented in the preceding chapter (and summarized in figure 5.4) is general enough to be consistent even with a position in which long-term choices were shown to be vacuously irrelevant in the face of dominant structural forces, whether physical, economic, social or political. Whilst this is technically true, it not only raises a myriad of basic philosophical questions (to some of which we shall return later). It also apparently demolishes any attempt at investigating urban dynamics since it raises the blanket-like causal mechanism of social structure to such an overarching position that it becomes impossible to identify any potential source of change.

Short of yet another, probably inconclusive, discourse in the philosophy of science, the only way of resolving such a debate appears to be through a close examination of the ways in which choice studies have actually been designed, of the illogicalities, faults and advantages which each approach appears to exhibit, and of the explanatory success – or lack of it – that each has achieved. The purpose of the remainder of this chapter is to review each of the major approaches with a view to shedding at least a little light upon the above proposition, and thereby more generally upon the overall question of the analysis of urban dynamics. For this purpose we shall concentrate exclusively upon the locational choices and decisions of individuals (whether acting as employers, employees, householders, etc.) rather than other long-term choices which need have no immediate spatial ramifications. The reasons for this limitation are, first, that such decisions are of obvious intrinsic relevance both to the interests of human geographers and to the needs of planning analysts. Furthermore, in developing a critique of the various analytical approaches – and in building constructively upon this critique – no special status will be accorded to locational and spatial variables, so that there is at least a good chance that such techniques as are found to be appropriate will also be readily transferable.

Since this is part of a mainly methodological work, we shall approach the review of location choice studies, not employing the usual land use/activity type of classification (manufacturing, offices, residence, etc.). Instead we shall attempt an initial typology of approaches which focuses upon the general theoretical framework and overall analytical style which is being or has been brought to bear upon particular relocation research questions. This seems reasonable, first, because it is a strategy which is more likely to yield generalizable conclusions, and second, because theoretical themes and methodological strategies do appear to have straddled the typological boundaries between the different types of location problem within this general research area.

More specifically, the literature yields at least three quite different approaches. The first and oldest of these approaches is undoubtedly that which derives directly from the application of the assumptions and techniques of neoclassical economics to the problems of location theory. In the 1960s and 1970s an accumulation of empirical assaults upon this approach brought forth a general diffusion of academic effort which suggested to several researchers that a less inhibiting theoretical framework might be more fruitful. Though it is more difficult to classify this phase of much more inductive theorizing, the general systems model from Bertalanffy (1972) and the biological sciences is probably as good an umbrella as any other. Finally, and as far as location analysis is concerned most recently, there has been a return to the theoretical level. This has taken the form of a shift in the treatment of locational issues back to the tradition of historical materialism. This work, whilst it has demonstrated very little interest in the process of locational choice *per se*, has had much to say about its outcomes, and so deserves attention in this chapter.

Neoclassical approaches

The neoclassical economic model has had a profound influence upon the whole area of locational analysis in planning and human geography. However, clearly the greatest influence has been in the two important areas of industrial and residential location theory. With respect to the key decision agents – the entrepreneur and the householder – the basic premises have been, of necessity, very similar in each case. Economic rationality has been assumed so that each has been vested with an absolutely complete information base, each with an optimizing goal of one sort or another, and each with the ability to rank all the options consistently in accordance with that goal.

However, despite this fundamental similarity, the theories proposed and techniques developed derive from quite independent academic traditions, though in each case the tradition is a long and richly exploited one. In the case of industrial location theory the pedigree can be traced back to Launhardt (1885) and subsequently through the works of such important figures as Weber (1909) and Losch (1940). Though different in crucial respects, these and most other neoclassical contributions to industrial location theory all spring more or less directly from the basic tenets of the economic theory of the firm. Residential location analysis, by contrast, springs from land market theory which goes back to von Thunen (1826). The best-known versions of the theory applied to urban residential land markets are undoubtedly those of Alonso (1964) and Muth (1969). Moreover, unlike neoclassical industrial location theory, the traditional urban economics approach to housing location is still being actively pursued (see for example Buttler 1981).

It is beyond both the scope and terms of reference of this

chapter to attempt a comprehensive review of work in these areas. Traditional industrial location theory has in any case been comprehensively and sympathetically reviewed by Smith (1971), and Evans (1973) has provided an equally detailed, and equally sympathetic, account of neoclassical residential theory. What we shall do is simply to outline the main thrust of each approach with a view to exploring, first, the practical analytical tools which have been developed in each major area, and second, the micro level postulates upon which the theory and the associated techniques depend.

Neoclassical analysis: assumptions and methods

Industrial location theory really starts with Alfred Weber (1909). He assumed perfect competition amongst producers whose markets and sources of inputs were all located at known fixed points. Thus within this competitive framework, characterized by exogenously determined prices and demand levels, the only rational course was to locate so as to minimize total costs, and these varied from place to place largely as a result of the varying costs of transport.

Realizing that the introduction of space inevitably infringed perfect competition assumptions, Losch (1940) proposed a monopolistic competition framework in which within a firm's market area it was the sole supplier of a given product. In the simplest form of the model costs and demand were assumed to be uniform over space, so that the goal of the optimizing firm had simply to become that of maximizing its market share, and this in turn was defined purely on the basis of spatially varying costs of product delivery. The final member of the group generally regarded as the founding fathers of the discipline was Hotelling (1929). His chief contribution was to introduce a semi-dynamic element into the study of location behaviour by introducing the notion of competitive relocation. The simple Weberian and Loschian frameworks assume instantaneous changes and the absence of locational repercussions.

Particularly during the 1950s and 1960s the pioneering work of these authors inspired a great deal of research activity. Much of the effort was directed towards theoretical elaboration through the progressive relaxation of assumptions, but this was frequently allied to more practical work intended to produce operational techniques. In the early years two areas of applied mathematics – namely those of mathematical programming and game theory – were stretched close to their limits in the cause of producing operational location models. Amongst the better known examples of attempts at programming models were those of Beckmann and Marshak (1955), Koopmans and Beckmann (1957), Goldman (1958) and of course the famous Kuhn and Kuenne (1962) heuristic programming solution to the Weber problem. The potential of game theoretic formulations to accommodate the phenomenon of competitive relocation was first explored by Stevens (1961) and then subsequently by Isard and Smith (1967). Unfortunately, despite the enthusiasm of its participants, and the ingenuity of their effort, the movement to operationalize industrial location theory ultimately failed. Many reasons have been offered and to some of these we shall return later in this chapter. It is worth noting here, however, that possibly the most recent attempt to offer a practically relevant planning tool which is strictly within the neoclassical tradition is that of Smith (1966, 1971), and his work goes right back to the basic essentials of the Weberian model to provide a simple graphic device for purely indicative locational analysis.

Academic history has been somewhat kinder to the traditional residential model. There has, for a start, only ever been one main point of theoretical origin (von Thunen), and though in the late 1950s and 1960s the application of his basic model to an urban residential context was pursued independently by several researchers (notably Wingo 1961, Alonso 1964 and Muth 1969), there has always been a far greater coherence within the overall programme of theoretical development than was evident within traditional industrial location theory. These original approaches share in common a view of the urban land market in which housing demand is made effective by individual households, all maximizing their own utility functions by trading off transport costs against land costs or rent. Some authors treat supply explicitly. Others do not.

However, most adopt a model in which an aggregate description of an equilibrated urban land market – or rent/price surface – is directly derivable from the micro level assumptions about the bases and goals of residential choice. This enables the derivation of a series of corollaries to the central theoretical premises. Perhaps the most frequently quoted of these is that which describes the equilibrium spatial distribution of households in different income groups, with the rich occupying the largest and cheapest suburban plots and the poor 'trapped' in the inner areas, occupying the smallest amount of space and paying the highest unit rents.

As noted above, ever since the work of the original land market theorists, research developments in this area have been keenly pursued. Right from the outset attention has been given to the problems of building operational models. One of the earliest attempts, and perhaps still the most famous, was that of Herbert and Stevens (1960) who developed a linear programming model whose primal involved maximizing aggregate household rent paying ability and whose dual yielded a rent surface for the whole study area. This model, though the closest parallel to the original theoretical work of Alonso, was never fully operationalized. John Kain's NBER urban simulation model (see Ingram *et al.* 1972) has, by contrast, been applied in at least two major empirical studies (both American). Though more complex than the model of Herbert and Stevens, the NBER model derives clearly from the same academic tradition – one in which households are assumed to choose housing types and locations that maximize their utility, and the market in housing is assumed to be perpetually moving towards a position of long-run equilibrium (though in the NBER case never actually reaching it). Even the work of Wilson and his colleagues (see Wilson 1974) in disaggregating the gravity type of spatial interaction model for the purpose of describing residential patterns, is in many ways consistent with the theoretical position of urban economists like Alonso. Their satisficing models, just as the optimizing version of Herbert and Stevens, describe access to work places as the key determinant of the residual housing budget and thus of residential choice. Moreover, several authors, including Senior and Wilson (1974), have demonstrated the parallels that exist between the two independent model development programmes.

There are other examples of operational models in this area. However, much more effort has undoubtedly been invested in purely theoretical development – to such an extent that Richardson (1977) has recently devoted the largest part of a fairly substantial book to a review of this work which he has labelled the 'new urban economics'. It would clearly be both inappropriate and impractical to attempt another review here.

Suffice to say that the bulk of this effort has been directed towards introducing sophistications to the basic model so that frequently observed and often discussed features of the urban land market may be treated endogenously. Included amongst these sophistications have been the treatment of extended household utility functions, relaxation of the monocentric city assumption, the introduction of externalities such as congestion and amenity variation, and even one or two attempts at introducing planning and the local and national public sectors more generally.

This programme of theoretical work has been paralleled (rather than precisely matched) by an equally extensive programme of empirical testing. Most of the latter has been aimed at exploring fairly simple versions of the basic model – thus the lack of a one-to-one matching of theory to test – but the sheer extent of the work is further evidence of the sustained interest in neoclassical residential location and housing market theory. Paradoxically, the equally extensive body of empirical work in industrial location research is evidence of precisely the opposite response, namely disillusionment with the basic theoretical approaches of the neoclassical school, for it takes the form not of a series of tests of the basic theories but rather of an inductive and open-ended search for an alternative.

Neoclassical analysis: limitations

In attempting to assess the neoclassical approaches to the problem of understanding locational decisions, it is pertinent to reflect upon this paradox, for to judge from a superficial review of the literature, its application in the sphere of residential location analysis appears to have been far more successful than it has been in industrial location studies. The most obvious place to search for the source of this difference is in the relevance of the fundamental behavioural premises. Though the same sort of economic rationality is assumed of the decision-making agents in each case, the paradigm might at least conceivably be a closer approximation to the reality of the choice process in the simpler residential case. In fact, there is no evidence that this is really the case.

The evidence – at least with respect to the behavioural assumptions – comes in both cases from studies which almost without exception take it as given that the traditional neoclassical theories are irrelevant. In other words the studies are not designed as tests of the original theories but rather as independent and inductively autonomous investigations or as tests of some alternative theoretical position. A recent review of industrial location studies (Brown 1979) expressed concern at this absence of direct empirical tests (specifically of the Weberian model). However, it is not difficult to understand why studies of manufacturing location decisions have not provided direct tests. The problem is that there is now ample evidence from mainstream economics of the prevalence of market imperfections such as prohibitive relocation costs, extensive information barriers and high levels of risk and uncertainty. The context of each decision is thus so far removed from that of the Weberian model that it becomes immensely difficult to judge the economic rationality of the stated intentions of investigated firms. Even more important, however, has been the discovery of the relative lack of importance of transport costs within the total cost structure of industry (see Chisholm 1971). In such circumstances it seems hardly worth persevering with Weberian techniques, dependent as they are upon perfect competition assumptions and emphasizing as they do the decisive importance of spatially varying transportation costs.

In this respect, however, there appears to be little to choose between neoclassical approaches to either industrial or residential location problems. For similar reasons residential choice studies (unlike empirical housing market studies) have generally ignored the simple micro economic theories. And those that have examined search procedures, priorities and outcome choices, have generally found it difficult to comment upon the overall rationality of the whole process, though they have discovered priority structures to be far more complex than is suggested by a simple trade-off function. Generally search patterns appear to be fairly localized (see Ford and Smith 1981) and to conform to a low-information satisficing model (Barrett 1976). Moreover access to work figures fairly low amongst the site selection motives (Stegman 1969) acknowledged by potential and recent movers, as marginally relevant in fact as are transportation costs to the mobile industrial plant.

If there is no difference in the relevance of their behavioural parameters, then the source of the difference in popularity between industrial and residential location theory must be sought elsewhere. An obvious second possibility is that the treatment of the context of the choice may be more acceptable in the residential case. In other words it may be that the assumption of a perfectly competitive housing market is more acceptable than that of a perfectly competitive market in manufactured goods. It may be further that the comprehensively coherent utility function is a more appropriate domestic model than a linear homogeneous production function is as a model of the average modern industrial plant.

It is, of course, patently clear that the contexts of locational choice are not as they were characterized in the original versions of either of the neoclassical approaches. Land and commodity markets are far from perfect, production functions far from linear and utility functions far from comprehensive. And yet all this would be freely acknowledged by most economists working today upon location theory issues. Much of their work is indeed directed towards extending and refining the original models so as to allow for imperfections and distortions of various sorts. The central point is that over the past few years the fundamental contextual criticism of neoclassical location theory has been pitched at an ideological or epistemological level rather than concerning itself with the detail. Thus the works of Massey (1974, 1978) and Storper (1981) on industrial location theory and Harvey (1973), Castells (1976) and Roweis and Scott (1978) on residential location and land market theory, whilst occasionally discussing particular explanatory inadequacies, generally mount their attack by first attempting to demolish the neoclassical framework *in toto*. And the most forceful argument applied to this purpose, certainly with respect to the neoclassical treatment of decision contexts, is that these contexts are drawn as timeless abstractions, independent of the realities of social, political and historical process.

Thus in order to derive stable theoretical pictures of those contexts, neoclassical theorists are forced into a particularly sterile methodological individualism. A set of uniformly anodized and dehumanized 'tastes' or 'preferences' have to be postulated as the ultimate and continuing sources of the market and locational patterns which are analytically derived. History does not enter the analysis and so the historical specificity of particular market relations and locational patterns cannot constitute parts of their explication. Factors of production become simply neutral technical inputs to abstract and formalized economic processes, rather than the products themselves of basic social and historical relationships. And the state – being manifestly the product of its own political history – has to be either ignored or at best treated as independent of the sphere of economic activity, perhaps acting upon it and responding to it but in no sense constituting a part of its overall structure.

We shall return at a later point to the alternative analyses of locational decisions offered from a structuralist or materialist perspective. The point worth emphasizing here is that each of the main streams of neoclassical locational analysis are equally open to this form of critique. Moreover, in practice, it was the von Thunen rather than the Weberian tradition which was first subjected to the critique (in the work of Harvey 1973) and subsequently housing and land market research has been more thoroughly scrutinized from a materialist perspective. The explanation for the continuing buoyancy of a neoclassical approach, at least in the residential sector has, therefore, still to be found.

Neoclassical analysis: applications

The simplest explanation, of course, will always be the one which points directly to differences in the pertinence of the theoretical accounts offered. If Alonso's theory is consistent with what we can observe and measure whilst traditional industrial location theory is not, then we need look no further. On the face of it, and given the above discussion of the behavioural premises and context descriptions offered in each case, one should not expect significant differences. However, the research evidence does suggest that residential theory may have the descriptive edge. The problem with neoclassical industrial location theory, and the source of its ultimate sterility, is that as it stands at present it is an adequate model neither of locational choice nor of the way in which aggregations of those choices constitute a spatial system.

Such interest as has recently been expressed in traditional industrial location theory has by and large favoured a Weberian model (Brown 1979). Thus of the alternative approaches available, those which focus upon supply side costs – the transport of materials and intermediate outputs, access to labour etc. – appear most relevant. The increasing size and complexity of market relationships have conspired to reduce drastically the relevance of market area models. However, the relevance even of Weberian analysis is called seriously into question as the balance in cost structure shifts further against transportation components. It is certainly clear that actual locational decisions rarely if ever seek transport cost minimization, so that Smith's approach which involves recasting traditional theory as an analysis of the framework within which location decisions are made appears to be the very most that may be salvaged.

All this, of course, is equally true of residential location theory. Though most of the components of the theory's utility function are no doubt amongst the factors that a house seeker must consider, there is no sense in which that theory may be regarded as providing a model of the procedures and outcomes of the process of choice. The crucial difference, however, lies in the comprehensiveness of the treatment of the decision context. Whereas Weberian theory has simply to assume the existence of a land market which determines rent and price levels, the von Thunen derivatives produce a description of such a market as a corollary of the micro level assumptions that they employ. This does not, of course, guarantee the accuracy of the description. As we have noted, the materialist critiques have been directed more forcefully at neoclassical land market analysis than at any other particular target. However, it does mean that perhaps the most critical of the artefacts with which planners have to cope, and in which human geographers are interested, is described explicitly. And it means, furthermore, that the theory, unlike the Weberian model, yields categorical propositions which are eminently testable.

The consequence has been that the so-called new urban economics has spawned, not only a rich and varied theoretical literature, but also a plethora of empirical studies of the urban housing market *per se*. A review by Ball (1973) discussed a large number of such studies, but interest has not abated since then – witness the work by Evans (1973), Richardson *et al.* (1975), Kain and Quigley (1975), Ball and Kirwan (1977), Wheaton (1977) and Diamond (1980). Virtually all of these studies (with the possible exception of that of Evans) end up equivocal on the question of the price-distance relationship. Most view a simple trade-off model as too gross an oversimplification. However, collectively they all bear witness to the continuing interest in the neoclassical model. Apparently for many it remains a useful way of analysing the critical phenomenon of the urban land market without requiring an empirically grounded account of the decision processes that constitute that market. What, then, is the current practical utility of the neoclassical tradition in industrial and residential location analysis? If any remains, this utility certainly does not lie in the framework's ability to generate formal and predictively powerful models of the outcomes of locational choices. The behavioural assumptions are just too wildly unrealistic for such a purpose.

Its residual value, therefore, must derive from the framework's ability, as an ideal type, to trigger insights which, whilst falling far short of comprehensive or empirically verifiable explanations, none the less capture something of the context within which locational decision processes are channelled. Thus, though no one would try to predict plant location choice on the basis of Weberian analysis, that analysis still illuminates, more effectively than most, the phenomenon of industrial and commercial agglomeration. Similarly, though no one would pretend that a neoclassical account offered much of an explanation of residential choice, it is still capable of shedding interesting and relevant light upon features of urban land markets such as segmentation and locational poverty traps. At the very least it provides a base line against which market imperfections and distortions can be assessed.

To return emphatically, however, to the central subject of this chapter, whatever the popularity – and limited descriptive success – of neoclassical housing market theory, as a micro level theory of choice it is as sterile as is traditional industrial location theory. The paramount importance accorded to travel and transport costs, and the continued reliance upon pure rationality and perfect knowledge assumptions have been specifically discussed above. More generally, however, the failure is probably best explained by reference back to the fundamental predictive purpose of positive economics – a purpose which was eagerly accepted by the location theorists of the 1950s and 1960s and indeed still is amongst schools, such as that of the new urban economists referred to above, which continue to thrive upon a diet of basically neoclassical assumptions. For those who have relaxed the fundamental behavioural propositions – as opposed to increasing the circumstantial complexity of their models – have found that the net result is predictive indeterminacy. And by and large, this is regarded as an unacceptable outcome. Illumination and insight are not enough.

Systems approaches

The predictive inadequacy of models based upon less restrictive behavioural assumptions has not meant that such assumptions have not been explored. However, these explorations have been for the most part reactive, attempting to overcome the deficiencies of neoclassical theory in an *ad hoc* and piecemeal fashion. A body of work has thus emerged primarily as a result of the perceived inadequacy of the building blocks of neoclassical theory. Few have denied the elegance of the theoretical constructs but many have been aware, at least since the worry was so concisely expressed by Pred (1967), of the enormous price of that elegance in terms of the mechanization of the fundamental model of goal-directed behaviour. The failure of Pred's own project demonstrated in the clearest of terms just how dependent those constructs were upon the artificiality of their micro level presuppositions. And the inescapable necessity of this dependence relationship has been evidenced repeatedly in the diffuse and inconclusive programme of studies which has been spawned by essentially the same worry as that of Pred.

This negatively motivated response is often labelled 'behavioural theory' (by for instance Townroe 1972, Wolpert 1965 and Thorngren 1970). Such a title is unfortunate both because its products do not constitute an alternative theory and because its objects are generally not (with one or two notable exceptions), the centrally dynamic processes of complex locational choice. More commonly the focus is upon the reaction of the firm or the household to an environment which triggers and then channels its choices. The process of choice becomes a black-box about which speculations are sometimes offered as a result of observing its outcomes (migration, relocation, plant rationalization, etc.) and its immediate or proximate environments (for instance its socio-economic context of domestic or organizational structure, and its information context of uncertainty or probabilistic knowledge). In other words, something akin to a systems model approximates most closely to the approach adopted by a large fraction of the studies in this area. Moreover, the looseness of this model as a guiding framework has simply reinforced the fragmentation of effort consequent upon the negativity of the original stimulus.

In this section we shall review some of the products of this general approach to location analysis. The lack of a tight theoretical core makes it even more difficult to distil widely shared assumptions and interpretations in this instance. Moreover, superficially at least, there appears to be little or no common ground between approaches to the location choices of households on the one hand, and those directed at the decisions of firms and other organizations in the products and service sectors on the other. It seems therefore appropriate to consider each of these broad sectors separately, at least to start with.

Systems analysis: assumptions and methods

Together the industrial and tertiary sectors have attracted an enormous investment of research effort over the past decade, and most of it may be not unreasonably thought of as adopting a systems approach to locational questions. It is certainly true to say that the overall approach, whatever we choose to call it, has taken the form of a reaction to the inadequacies of neoclassical theory, rather than that of an independently motivated academic venture built upon its own unique and internally coherent intellectual tradition. The central assumptions have therefore been, for the most part, negative ones. First and perhaps foremost, systems approaches invariably assume that individual decision-makers do not make locational and complex investment decisions so as to maximize profits. The fundamental behavioural postulate is generally far closer to the model which Herbert Simon (1959) labelled 'satisficing', and so typically emphasizes the achievement of targets or standards. A necessary corollary of this reactive assumption is the proposition that, just as the goal of complex decisions is not that of optimization, so the context of those decisions is not that of a competitively integrated set of neatly self-equilibrating market places. A precise contextual postulate is rather hard to pin down (though generally accorded critical importance) but it frequently focuses upon the complexity of institutional relationships and dependencies which are neither purely economic nor effectively self-regulating.

Though differing widely with respect both to technique and to conclusions, an emphasis upon different levels and types of decision context has been a fairly widely shared theme amongst the empirical studies. More specifically most research has addressed one or other of two broadly distinguishable types of context. First, many studies, especially in recent years, have laid considerable emphasis upon the long-term macro-economic and technological contexts of location decisions, addressing issues such as declining aggregate demand and the automation of production. Others have focused more specifically upon the more immediate context of location adjustment represented by the organizational structure of the decision-making unit itself (the firm or institution).

A long-term historical perspective is normally associated these days with a materialist approach to spatial analysis. However, many studies adopting a systems perspective start by specifying the relationship between the long-term dynamics of the parent economy and medium-term adjustment strategies of industrial and other economic activity. Thus frequent reference is made to the impact of technological innovation upon the product cycle. Building upon the work of Galbraith (1967), analysts such as Rees (1979), Gilmour (1974) and Tornquist (1977) have explored the locational implications of this process. Factors such as an accelerated rate of plant obsolescence, the increasing complexity of inter-industry linkages (through for instance component specialization), and of course the reduction in the relative significance of transport costs, have all been much discussed. Their combined impact upon the process of locational adjustment has never been specified formally, but their collective importance is generally acknowledged.

One of the most frequently discussed corollaries of these trends has been the shift in the structure of modern economies away from a predominance of manufacturing employment towards an aggregate preponderance of the service sectors. Indeed it is this particular trend which now most frequently constitutes the central plank of accounts of inner city malaise (see for instance Department of the Environment 1977 for the most recent official analysis in this country and Hall 1981 for a concise review of recent academic thinking). It is argued, to put it at its most simple level, that the problems of the older inner areas of large cities stem ultimately from the exodus of employment. The economies of these areas were built firstly upon large-scale heavy engineering, food processing and the break-of-bulk in raw materials, and secondly upon small-scale labour intensive trades such as printing, tailoring and the manufacture of furniture. The restructuring of industry in the face of new technologies and a shifting and declining pattern of aggregate demand has hit the inner city hardest because its large-scale plants are old and use their capital inefficiently and its small-scale units are the sort which, because of their sub-contractual status and inadequate reserves, always go under first when times get hard. Thus through branch plant rationalization and small firm closures, the inner areas have lost an alarming number of their manufacturing jobs over the past fifteen or twenty years.

However, if this pattern of localized decline is explicable in terms of overall technological trends, what have systems approaches to offer by way of an account of the location choices of employers in the tertiary sectors? After all, the inner city has traditionally been the home of these sectors and so the aggregate shift in their favour, so often mentioned in systems accounts, should have provided compensating increases in labour demand, more than sufficient to offset the declining manufacturing base.

In fact, analysis of tertiary sector location choices is normally pitched, in systems models such as those of Thorngren (1970), at the second of the contextual levels mentioned above. It is the structure of the organization from which the decision emerges, rather than that of the parent economy, which is treated as critical in determining spatial outcomes. Studies typically start with the proposition that inputs and outputs predominantly take the form of written and verbal communications which exhibit complex spatial cost surfaces. They then move on to investigate the importance of this fact in determining the resolution of locational questions. Thus, over the past twenty years there have been a whole series of relocation studies designed generally to explore the criteria treated as critical by the organizations involved in spatial adjustment. Many factors have emerged in these studies, including investment and land cost differentials, effects upon the organizations' image or prestige, the ease of retaining or recruiting lower grades of staff and even a bias in favour of risk aversion. However, by far the most important of the factors to emerge have been, first, the need to retain senior and specialist staff, and second, the desire to preserve or improve internal and external communications structures. And of these two, the latter has been the most frequently mentioned of all locational criteria.

Of course this does not mean that the overall economic context has been studiously ignored. Most studies acknowledge the importance of shifts in aggregate demand, at least for the purposes of explaining the sources of tendencies towards spatial upheaval. The point is simply that most of the work was done during a period of continuing economic expansion in the tertiary sectors, and any analysis of this backcloth came to be regarded as a commonplace. Thus the most interesting theoretical work (that of Thorngren 1970), and the enormous body of empirical work which it spawned, took not only this economic backcloth but also the pivotal role of communications structure as given elements in our models of tertiary sector locational choices.

Thorngren's project therefore became a highly specific one. His aim was to classify the internal and external communications of an institution in terms of their organizational level and functional role. In this way he hoped to be able to delineate its most appropriate environment of contact opportunities, and thus also its most likely locational choice. He ended up with a three-way classification of linkages – orientation, planning and programming contacts – in which each successively lower level was held to involve smaller groups, lower levels of seniority, higher levels of routinization, and thus more closely bounded contact environments. The point was that each level required a clearly specifiable and different balance between information overload and starvation, and, as distinct groups of tertiary sector firms and institutions exhibited differing degrees of dependence upon each of the three levels, such groups would probably also exhibit differing locational requirements.

Such analysis clearly raises many empirical questions. The most obvious of these is whether or not tertiary sector organizational structures do indeed exhibit such communication patterns, and if so, whether their locational choices reflect significant differences in these patterns? Many studies have been designed to address these and related questions. The relevance of the Thorngren model for describing something of the communications structure of the organization has been demonstrated by several studies (see for instance Goddard and Morris 1976). The precise connections with locational decision-making are more problematic, but there is at least indicative evidence of the spatial clustering of closely related organizations in the tertiary sector (see Goddard 1973).

More generally, the study of organizational context has become an increasingly important component of the systems approach to locational investigations in all sectors. It has become a commonplace, even in the study of industrial location, to preface analysis of locational choice with an exploration of the type of industrial organization from which the choice emerged. Thus in recent years, especially since it has become necessary to focus industrial location analysis upon the issues of rationalization and closure rather than upon those of expansion and development, attention has been directed mostly at the ways in which typical forms of the large-scale manufacturing enterprise respond to a changing economic climate. One issue that has been hotly pursued is that of the way in which multi-plant firms implement a programme of rationalization (see Townroe 1975 and Dicken 1976). This is clearly of considerable importance given the increasing dependence of many local economies upon satellite branches of such corporations. It has been suggested (Erickson 1980) that a three-way typology of firms – differentiating vertically integrated, horizontally integrated and single product organizations – may be of explanatory value in that each class exhibits a different pattern of branch plant interdependence. However, research findings in this area remain equivocal.

Though the issues are very different and the operational research strategies far removed from those employed in the studies of industrial and office location, the emphasis upon response to context is still clearly discernible in the rather smaller set of systems studies of residential location. It is even possible to apply something similar to the two-way classification of contexts used above. For, though the sheer volume of research in this area is much smaller, several studies have addressed the problem by differentiating between the external environment against and within which the choice is made and the attributes of the household or family unit which makes the choice. The distinction between generalized and specific contexts typically found in residential choice studies is, however, a much simpler one than that which characterized the systems approach to industrial and office location analysis. There the boundaries between the organization on the one hand and the economy and its technology on the other were only broadly defined and were treated differently from study to study. Most of the so-called systems approaches to housing choice share a common basic model which accords a clear role to each of the two decision contexts. Moreover most also share a broadly similar understanding as to the components of each of the contexts.

The model referred to is an essentially behaviourist formulation based upon the notion of stimulus and response. The generalized context is said to encompass a variety of attributes of the physical environment. And the specific context comprises attributes of the family such as stages in the life cycle, age, size and income. Thus most studies build upon the early work of Rossi (1955) and Wolpert (1965) and regard the residential environment as a continuous source of stimuli. The household's response to this collective onslaught is mediated by evaluative filters which are in most models assumed to be related in a fairly simple fashion to the sorts of socio-economic variables mentioned above. Stress is said to result therefore either from a change in the residential environment or from a change in the household's needs or aspirations which has the effect of changing the way in which the environment is evaluated. And 'stress' (in Wolpert's rather idiosyncratic use of the term) is the source of a decision to move. There have been several studies over the past ten years which have adopted and explored this overall approach (see for example Brown and Moore 1970, Herbert 1973, Speare *et al.* 1974, Short 1978 and Coupe and Morgan 1981). The outcome of all this effort has been to demonstrate several significant relationships between household contextual variables (particularly life cycle variables) and the decision to move. However to isolate patterns of concomitant variation is not to give a full account of complex social choices such as are involved in a migration decision, and it is probably fair to say that researchers are nowadays agreed that the stress-response model offers far too simple and mechanical an account of the emergence of the decision to move. Some studies have attempted to control for personality variation (for instance that of Salling and Harvey 1981) but the majority simply attempt to link the decision to move in a unidirectional model to one or more 'external' stimuli, whether generalized (i.e. observable attributes of the physical environment) or specific (i.e. observable attributes of the household).

Systems analysis: limitations

Perhaps the central problem with systems accounts of industrial, office and residential location decisions is that they are not models of the decision process at all. Dissatisfaction with the behavioural assumptions of neoclassical theory has certainly prompted a wealth of innovative research. Moreover this research has adopted a much more flexible overall framework (the systems model) than that of neoclassical work (the self-equilibrating land, property and product markets). And of course it has dropped the basic behavioural assumptions of utility and profit maximization in favour of looser models such as the satisficing model of Simon (1959) or the stress-response model of Wolpert (1965). The point is, however, that each of these alternative behavioural bases is adopted, when it is adopted, as a fundamental premise rather than an hypothesis to be tested.

The systems framework actually reinforces this tendency since it encourages the analyst to explore the transformation properties of a black-box simply by examining the variety of the relationships between its inputs and outputs. It offers little help to the researcher who wishes to examine the processes of the black-box directly, and none at all if such an examination is to allow non-mechanical accounts of these processes. Thus, as we have noted above, the vast majority of the systems approaches to locational decisions avoid any direct treatment of the decision process itself in favour of indirect analysis which relates decision contexts to outcomes (whether these be specific spatial choices or simply the decision to make a move). In other words, this context-outcome or stimulus-response formulation, so characteristic of work in the field, is a formulation which follows quite naturally from the mechanical approach to theorizing and generalization which lies at the heart of the systems analysis model itself.

One of the difficulties which systems approaches must face as a result of this indifference to the actual process of choice is that of explaining urban dynamics. It is clearly the case that the 'urban system' changes over time both as a spatial and as a social entity. Part of that change must spring from the cumulating impact of location choices. Yet if the outcome of the location choice black-boxes remains a mechanical function of its input then we must look to this input for the source of urban change. The problem is that as a systems model increases in comprehensiveness so the source of its dynamic disappears into the analytical distance. Ultimately the source of all change is normally traced back through a series of functional relationships to the interface between environment and technology which is therefore by definition inexplicable in systems terms.

It is not that change cannot be comprehended within a systems framework. It is rather that it cannot be generated by any of the structures or relationships which that framework describes. The analytical tools of the approach do not stretch that far. They illuminate by means of fragmentation. Relational structure is thus contained in the pattern of input-output, negative feedback or other dependence relationships which enmesh the subsystems. The forces of change are thus externalized. The structural transformation of a system is ultimately explicable only in terms of that system's environment. The system is, in the end, nothing but its networked form and so the only place to look for its transformation is on the outside.

If the systems model is too narrow to encompass its own dynamics, it is, at the same time, too broad to stimulate a coherent line of theoretical development. Possibly because the movement started out as an empirical reaction to high theorizing, it has often tended to swing too far in the opposite direction of inductive opportunism. Thus generalizations built upon one sample survey as often as not bear no comparison with those built upon the next. Fortuitous correlations support ephemeral conclusions, and the overall picture becomes confused and sometimes even contradictory.

Systems analysis: applications

Systems studies have, nevertheless, served several useful purposes. They have, once and for all, dispelled any residual misconception that locational choice could be treated as the simple addition of a spatial dimension to neoclassical economic models. They have demonstrated the complexity of the relationship between context and locational outcomes. They have thus convinced academics of the difficulties of explaining locational decisions; planners of the difficulties of attracting and holding employment generating activity; and entrepreneurs of the difficulties of making investment decisions.

Their influence in practice has, moreover, been direct as well as indirect, at least in the field of industrial location decisions (see Keeble 1976). Planning departments in the UK are almost all now facing the same central problem of a declining economic base. They are all competing, with the most limited of tangible incentives, for such footloose job creating enterprise as there is, as well as desperately attempting to hold on to the firms and institutions which have so far survived. One answer that has occurred to analysts in many of these departments is that what they lack is adequate information, and the response has, in many cases, been empirical research, in the tradition described above, into the locational and other environmental requirements of the firms in their areas. As the direction of policy initiatives has shifted, interest in such empirical work has not abated. Thus the newly formed local enterprise boards have all followed the lead taken by the Scottish Development Agency by developing detailed analyses – generally based upon surveys of small samples of firms – of locally important industrial sectors (fishing and footwear in Lancashire, furniture in London and the foundry and automotive sectors in the West Midlands).

Materialist approaches

Whilst adopting much of the methodology of the systems approaches of the 1970s, studies such as those in London and the West Midlands share none of the aspirations of their precursors. Most of the earlier systems studies espoused a centrally academic purpose. Their aim was to replace what their authors believed to be a sterile model of choice and context with one which was more realistic in a behavioural and contextual sense. When the research was bent to a practical purpose, its target quite often became that of advising private sector firms and institutions facing complex location decisions (see Townroe 1976). The practical problem addressed through research sponsored by the enterprise boards in London and the West Midlands is quite different. The problem they face now is one of deteriorating metropolitan economies and persistently high levels of under-employment.

Whether or not the escalating scale of this problem has been the source of a leftwards shift in urban political opinion, such a shift has occurred. Moreover, a similar shift has been apparent in the world of location theory. Indeed, historical materialism, as a new theoretical core to mainstream location theory, has probably become its central and dominant core over the last few years. In some superficial respects, this new movement shares a great deal in common with the systems model which it is attempting to replace, but in a more important sense it represents a departure in assumptions and method as fundamental as any so far encountered in this review.

Materialist analysis: assumptions and method

Like systems approaches, materialist analysis rejects the artificial behavioural basis of neoclassical economics. Like systems approaches, materialist analysis rejects the omission of an explicit treatment of the non-market context of location choices. And like systems approaches, materialist analysis relies very heavily upon the treatment of contextual factors in explaining the outcome of locational issues.

However, this is where the similarities end. Unlike systems approaches (which were originally devised for the study of biological phenomena), materialist analysis has always been a social scientific technique for the study of social phenomena. It is based upon the principles of historical materialism expounded originally by Marx in *Capital* (1867) and elaborated extensively since then, and it has always faced up to the problem of analysing social (and more recently urban) dynamics. The way the materialist approach tackles this latter problem is by introducing into the analytical armoury a relationship unused and unacknowledged in systems analysis – namely the dialectical relationship of fundamentally irreconcilable interests. It is argued that all real social change springs ultimately from basic inter-group conflict. Systems approaches fail, so the argument goes, by stressing the functional integration of urban institutions and by deriving the social relations of urbanism (in the tradition of Louis Wirth 1938) from an analysis of spatial form. Thus, according to this view, there can be no conflict relationships of which urban forms and institutions are mere manifestations. At most urban forms are described as having side-effects which are dysfunctional and cause patterns of behaviour which are labelled as pathological, deviant, insane or criminal (see Milgram 1972). They are, in other words, beyond the realm of the normal, functional social world.

Materialists working in the urban sphere of analysis generally accept the reinforcing and even alienating power of spatial forms. However, they argue further that both spatial configurations and the urban institutions they accommodate are fundamentally just manifestations of basic production relationships. The crucial relationships are those of ownership and exploitation, and they are, clearly and intrinsically, conflict-inducing.

On this simple premise is built a vast edifice of theoretical development and empirical elaboration, at the ends of which come specific studies which bear upon the issues of industrial location and urban residential land markets. It is acknowledged that there is as much variety within the category of materialist approaches as there is, for instance, within the systems group discussed above. It is beyond the scope of this book to do adequate justice to this variety, and so what is presented is a simplified account of urban processes which derives ultimately from Marx, but as interpreted by authors like Castells (1976), Harvey (1973, 1981) and Poulantzas (1975).

It may appear unreasonable, in a chapter devoted to studies of locational decision-making, to attempt even a simplified account of this work. If neoclassical and systems approaches only just qualify as theories of social choice, it is probably reasonable to say that materialist accounts do not qualify at all. Indeed most of these would probably reject altogether the idea that there could be any body of theory about industrial or residential choice and location which was independent of the historical and structural rules defining the development of industrial capital and the role of the state. However, we shall pursue the discussion a little further here partly because much of the current research which bears significantly upon industrial and residential location questions is firmly set within the materialist tradition, and partly because this research may not unreasonably be regarded as in some senses a logical extension, as well as a significant diversion, of the neoclassical and systems traditions.

We noted above that there were similarities between systems and materialist approaches, and that there remained a central difference which had to do with the analysis of change. Whereas the crucial dynamic in systems approaches (technological change) remains firmly external to the analysis, the dialectical method reinternalizes it as a fundamentally inconsistent and unstable framework of macro-economic circumstances to which each micro-economic decision is a repeated response. The ever-widening oscillations between expansion and recession, said to be intrinsic to the process of capital accumulation, entail technological innovation as just one defensive strategy amongst many others. These others include organizational restructuring, plant rationalization and relocation. Thus industrial and office location decisions become parts of a much wider economic context, just as they are in reality in neoclassical theory. The difference is that the neoclassical context is characterized by a pattern of functional integration (as in systems approaches but more narrowly economic in scope) whilst the materialist context is characterized by one of intrinsic conflict. The neoclassical context tends to restore equilibrium, the materialist one to destroy it.

There is still not a large number of empirical studies of industrial location set within this framework. However, Massey's (1978) work on the UK electronics industry does demonstrate clearly the way in which economic crisis and a falling rate of profit are translated down, through a variety of defensive attempts to raise labour productivity, to the necessity of adaptive strategies which have locational implications. These implications of course include rather dire ones for the older inner areas of large cities (Massey and Meegan 1978) in which are to be found the least productive units of capital. Thus by this route we are provided with an alternative account of inner city economic misfortunes, and perhaps one that is somewhat less tortuous than that provided within an essentially neoclassical framework.

Though the empirical studies have been few in number the treatment of industrial location within an essentially economic model of society such as that of classical Marxism presents few major theoretical problems. The same cannot be said of the treatment of residential location. The trouble is that in a mixed housing economy any analysis of residential location must take account of state involvement, both directly (through the non-market provision of accommodation) and indirectly (through planning and other controls over the residential land market). Such an extension would clearly be very difficult within a neoclassical framework though there have been one or two cautious, limited and by and large unsuccessful attempts to incorporate state controls within new urban economics models (see for instance Ohls *et al.* 1974). There has been, it would appear, virtually nothing at all within the systems tradition which treats residential choice in the public sector.

This does not, however, mean that, with respect to housing and residential location analysis, there remains no alternative to a materialistic perspective, suitably extended to reflect the role of the state. The institutional complexity of housing markets, coupled with the contextual *naïveté* of the neoclassical and systems studies, has spawned a large and diffuse literature, focusing generally upon the institutions of housing supply, sometimes from a so-called corporatist or managerialist perspective, but invariably stopping short of a fully fledged Marxist interpretation. This literature, reviewed in Bassett and Short (1980), covers ground far too extensive to treat here. Moreover, despite its breadth and descriptive interest, it is not centrally relevant to this discussion since it reflects neither a distinct theoretical position nor a coherent methodological base.

As we have seen already, historical materialism certainly suffers from neither of these deficiencies. However, the institutional complexity of the housing market, brought out clearly in the studies referred to above, does present difficulties for those working in a Marxist tradition. Perhaps the key problem for materialist analysis in planning and human geography is that alluded to above. Marx did not himself provide a fully worked model of the state on which to build a mixed economy housing market theory. Others, however have done so building upon the structuralist theories of Althusser (1969) and Poulantzas (1975), and it is upon their work that materialist studies that touch upon residential choice and location issues, for the most part, depend. At its simplest the neo-Marxist theory of the state accords the institution no more than instrumental status. The state, and thus also the local state, and within that context planning as a local state activity, all become instruments of the owning and exploiting class. They are indeed necessary extensions of the free enterprise system brought forth as a result of its inherent instability and tendency towards crisis. Their function is the manipulation of the artefacts of the system – the cities and towns – for the single and fundamental purpose of reproducing its one essential ingredient, namely relevantly trained and accessible labour power. Thus the apparent paradox of a capitalist system characterized by ever-increasing state involvement in transport, education, health care, welfare services, and of course housing, is resolved if each further encroachment is interpreted as yet another defensive response to falling rates of profit in the cause of increasing labour productivity. At least in the simpler models, the state is accorded no real autonomy at all.

Research within this framework, or something like it, of course has very little to say directly upon the issue of residential choice. However, there is now a significant body of work attempting to rethink housing market models replacing the neoclassical framework of Alonso (1964) and Muth (1969) with that of neo-Marxist economics. The most influential theoretical works are probably still those of Harvey (1973) and Castells (1976). However, since then more closely applied and sometimes empirically based studies such as those of Harvey (1977), Ball (1978) and Roweis and Scott (1978) have attempted to show in some detail how the various institutions of a capitalist economy – most obviously the national and local states but also the financial institutions (see for instance Boddy 1976) – shape, distort and segment spatial housing markets. Moreover, a conflict model such as the one outlined above also permits the analysis of housing topics like the squatting movement which would always defy treatment in neoclassical or systems terms. Thus Cant (1976), building upon Castells' (1976) work on urban social movements, attempted to characterize the squatting movement as a response to the state's increasing involvement in the housing sphere of distribution. Housing was thereby politicized, evoking as a response a squatting movement articulated along political lines, and precipitating an essentially unstable housing market relationship in which the state was inextricably involved.

Nothing is said of the motives and interpretations of the squatters themselves because nothing needs to be said. The model is fully comprehensive. The advance represented by the materialist approach might be summarized as follows. It has translated the economic relationships of neoclassical analysis into economic power relationships and thereby found a way of internally dynamizing its analysis without risking the dangers of infinite regression associated with a humanist approach to social change. To argue that change springs from deliberated human choice begs questions as to the causation of that choice. To reply that choice springs from motives and attitudes begs questions as to the causation of non-observable behaviour.... And so on. Much safer, therefore, to leave out the individual altogether and locate all explanations (including those of social dynamics) in the dialectical relationships that vitiate social structures and economic institutions.

Materialist analysis: limitations

We shall return to this argument in the next chapter. For the moment, it is worth considering briefly the price of accepting it and thus also of accepting a materialist analysis of the outcome of a complex social choice. Clearly such an analysis has a great deal to offer when compared with neoclassical and systems approaches. Neither of the latter cope effectively with the conflict that is so often associated with spatial choice. Neither offer models with internally driven sources of social change. And yet neither compensate for these omissions by analysing the process rather than simply the context and outcome of a social choice.

What then is sacrificed if we respond, as so many have responded in recent years, by accepting a materialist base as the only legitimate one for a programme of analysis which aims at something more than the aggregate statistical description and extrapolation which can be achieved using the crude techniques discussed in chapters 3 and 4? There appear to be at least two significant difficulties attached to such a commitment. The first is that as a contextual account of specific issues it is at once too narrow and too remote. It acknowledges the causal efficacy of only one social relationship, namely that of property, and so appears awkward and even irrelevant when applied to issues like nuclear defence and amenity preservation which seem to transcend ownership and class divisions. More particularly, since an historical materialist account cannot readily accommodate the localized relevance of locally specific features of the social or physical context, it offers no help to those who wish to investigate, for instance, why one local spatial issue was resolved in one way, and another, apparently very similar, in a quite different fashion. To say of every specific locational issue that it all boils down to the exploitation of labour by the property owning class is to say very little that is either of practical value to the hard pressed planning research department, or of explanatory relevance to the uncertain human geographer.

The second problem is that the structural formalism of the model may be regarded as demanding a cynicism about the social and political actors involved in a locational issue which is unacceptable. If, for instance, one believes that the agents of the state may be capable in some circumstances of acting in ways which are at least partially independent of their 'true' class interests, then such a belief appears very difficult to accommodate at least within a structuralist framework. It may be protested that a strict Marxist position is consistent with the assumption that individuals act upon genuinely altruistic motives. The point is simply that whatever their motives the organization of the society in which they operate will eventually ensure materially predictable effects – until, that is, their consciousness of this pattern of social organization is raised to a level sufficient to precipitate a revolutionary response.

In fact this is an inadequate answer to the charge of determinism since it preempts and thus trivializes the reflective autonomy of the individual. It presumes there to be only one sort of potentially effective consciousness, only one way in which that consciousness may be raised and only one way in which it may be converted into social change. We are now, however, back where we started. It appears that if we are to allow of any degree of individual autonomy we must also allow that social change may take a multitude of forms and may spring directly from the deliberated choices of self-consciously reflective individuals. None of the approaches reviewed in this chapter seem capable of accommodating such a premise, and so none of them take us very far beyond the position reached at the end of the preceding one. There we reached the conclusion that a study of social choice as a coherently structured process appeared to be essential if we were to approach an understanding of its outcomes and thus, more generally, of the dynamics of urban social life. Here we have found that there is an alternative route to genuinely dynamic analysis, but one which will never satisfy those who find dialectical mechanism as limited a basis for social theory as functional mechanism is now for the new generation of materialist analysts in planning and human geography.

Materialist analysis: applications

This generation has yet to demonstrate the practical relevance of their work, but would probably find the idea that it should be required to do so at once both amazing and unacceptable. In one sense materialist analysis is by definition of practical relevance in the political sphere because it is intrinsically political. It is rooted in political and economic history and is incomplete if it fails to reach political conclusions. Thus Massey and Meegan (1982) conclude their study of job loss both by attacking incentive-based regional policy as a way in which the relative power of capital is increased through job elimination, and by drawing the entirely logical conclusion that future growth will depend more upon political (power rebalancing) than upon technological change.

Materialist analysis of housing and residential structure has, as we have seen, been even more intimately concerned with the role of the state than its counterpart in the sphere of industrial location. As a result policy and political questions have always been that much more central. Much effort has been expended in attempts to explain the precise history of housing policy in Marxist terms (see for example Dickens 1977). Moreover, the detailed empirical scrutiny of closely defined industrial sectors made recently by Massey and Meegan (1979, 1982) was in a sense previewed in studies of the local organization of housing policy by Cockburn in 1977.

What is lacking in most of this research is evidence of practical applications. Up to this point the lack of such applications has usually been explicable in terms of statistical crudity, theoretical inadequacy or technical complexity, and it is undoubtedly the case that materialist studies can be found which suffer from each of these defects (even the first). However, the novelty of the Marxist tradition is that it offers only political, not policy, guidance. In other words, it rejects the presupposition that the products of its efforts should be expected to answer policy questions in the terms in which they are normally expressed. Materialist analysis is applicable in the sense that it is directed towards political change, not in the sense that it can assist in the task of context-given policy manipulation.

Epistemological problems

If this is an inadequate model of applied social science then its inadequacy does not follow from its own internal incoherence. The framework all hangs together beautifully. But then so does the neo-classical alternative. Each works in its own terms as a self-sufficient theoretical construct. Yet neither yields a satisfactory model of urban dynamics.

We have already indicated the probable reason for this apparently ubiquitous limitation. The problem that is shared by both theories – as it is with most of the systems approaches – is that these theories ultimately fail to grapple with a realistic model of the process of social choice. By reviewing specific location theories and studies we have shown that this has in practice generally been the case. It is possible, I believe, to go one step further and show that this tendency was more or less necessarily built into their theoretical starting points. Thus the crucial weakness of the intellectual frameworks considered in this chapter is quite simply that each, no matter what its context-defining virtues, builds upon a model of human choice which is reactive rather than creative; determined rather than determining; mechanical rather than reflexive. The interesting point is that each of the frameworks achieves the mechanization of its model of choice in a different way.

The neoclassical approach focuses upon the concept of optimization. Superficially, this appears to reflect a non-mechanical model of choice since it presupposes one or more goals on the part of the decision-maker. In fact the goals are, in at least one fundamental sense, context-defined. A market organization yields costs and prices and so goals are allowed to be no more than mechanical functions of these variables. As for the procedures of choice, these in their turn amount to no more than an exhaustive series of reactions to further aspects of this market context. A search through the information which constitutes the market – a search whose features neither are nor need to be discussed or described – is all that is involved in the process of neoclassical decision.

The position is not much better in the case of systems models. The crucial concept in this case is generally that of adaptation. The decision-maker is not assigned a single goal but a set of tolerance levels or choice thresholds. Each of these constitute what may be thought of as environmental filters. The mechanism of choice thus becomes a process of adaptive response: the context is tested against the relevant set of thresholds until a satisfactory location is found. The outcome represents an adaptation to circumstances which are fully defined by the relationship between the individual's externally given context and his or her predetermined threshold specifications. Of course, a non-deterministic model of the process of establishing tolerance levels could be imagined. The point is that systems approaches to location theory do not explore such possibilities (just as neoclassical analysis does not explore the process of goal formation and selection). And just as the neoclassical framework of integrated markets neither offers nor requires any non-mechanical treatment of goal formation, so the systems framework of self-adjusting subsystems neither offers nor requires such a treatment of adaptation levels.

Materialist analysis of locational change is at least different in that ostensibly it ignores the issue of choice altogether. Locational decisions occur none the less, and in that materialists purport to explain their outcomes, it is reasonable to note what the implicit or explicit interpretation of the decision process is or must be. In fact the concept central to this interpretation is also central to the overall social model, and this is the concept of exploitation. Context-defined class interests result in actions whose effects are to defend and reinforce the relationship of exploitation. At one level, therefore, the epistemological differences disappear. The model of structure defines the process of choice in much the same way as it does in both neoclassical and systems formulations. The only difference is that a structure which, because of its internal contradictions yields exploitative action, also yields as a dialectical response that of alienation, and so contains the seeds of its own transformation.

At this point in previous chapters I have attempted a sort of localized synthesis. This has generally taken the form of a selective recombination and reformulation of some of the techniques reviewed. It has been intended to act as a means of finding a level for those techniques which is not only realistic but is also theoretically and epistemologically appropriate. In the last chapter, however, the attempt was neither comprehensive nor particularly successful. Paradoxically, this was because that chapter started to examine techniques which raised central theoretical questions – questions for which there are no easy answers. What we have discovered in this chapter is that those questions literally cannot be addressed appropriately if we confine ourselves to the terms and the strategies characteristic of traditional methods. Moreover, this conclusion is as true of the methods of historical materialism as it is of those of neoclassicism. Synthesis, therefore, must be sought by extending the boundaries of the discussion.

7

Prediction, explanation and interpretation in applied urban analysis

If the last two chapters are thought of as working towards the common aim of providing techniques and strategies for exploring the process of urban dynamics, then clearly we still have a long way to go. Those quoted in each of these chapters, whether geographers or planners, would probably not accept that that was all or even part of their overall purpose, but none the less this inference of an ultimate target for their efforts seems not unreasonable. From the planner's point of view an understanding of the dynamics of urban systems is absolutely crucial. Chapter 5 demonstrated that studies of attitudes, of values and of daily lifestyles were at their weakest when treated as theoretically self-sufficient and thus as providing independently coherent and unambiguous models of social life. At their best they may constitute relevant building blocks for more ambitious studies of structured locational choice. In chapter 6, however, we observed that by and large such were not the building blocks upon which the currently popular models of locational choice relied in practice. Moreover, as a set these models, with the possible and notable exception of some materialist accounts, failed conspicuously to add to our knowledge and understanding of urban dynamics. This lack of success did not amount to an 'in principle' failure however. The theoretical proposition that social change results as context is self-consciously manipulated or dialectically disturbed through individual or group participation in the process of social choice was neither explored nor denied in these studies. For the most part it was simply ignored.

This oversight will be addressed later in this chapter. In a sense we shall then be continuing the argument of the preceding chapter by returning to the issue of structured locational choice, but a break at this stage for a more broad ranging overview of the earlier discussions will probably be helpful. The point is that, though what we have yet to cover is critical to the logic of the overall argument, as well as an extension of the themes most recently treated, it will be an inevitably speculative extension. Before we embark on such a programme of speculative elaboration it will be as well to clarify just what may be pertinently offered to the analyst in planning and human geography whose needs are far from completely satisfied by the techniques and approaches currently available. And the best way to do this is to review these techniques – the techniques discussed severally up to now – so that we may be clear about the functions that are adequately fulfilled and the investigational gaps that are still left uncovered.

The typology adopted to this point was designed to classify the techniques of analysis purely on the basis of the level of aggregation of the objects of the investigation. Though this served an initially useful organizing purpose, its utility was exhausted about half way through the last chapter when we observed that what started as a structural analysis of (perhaps) a single locational choice could end up as a discussion of the organization of a whole society. The explanation, in other words, is only fortuitously placed at the same level of aggregation as that which is explained. If we are to look at the relationship between function and method in applied social science, then a simple macro-micro dichotomy is likely to be too crude. Much better to focus upon the mode of analysis, the investigational purposes which that mode can support, and the basic method it suggests.

Though it may seem that the investigational purpose or function of any piece of analysis – whether this be prediction, explanation or interpretation – must be its most basic attribute, this is, in at least one important sense, not the case. The point is that there appear to be fundamental modes of, or approaches to, the problem of analysis which have a preemptive effect upon investigational functions. Certain sorts of tasks for which analysis may be required are just not achievable within given analysis modes. That this is infrequently acknowledged is probably due to a reluctance to push the functional requirements of analysis beyond the limits of its simplest mode. And this mode may perhaps be most appropriately labelled the mode of comparative statics. Such a mode is characterized by a focus upon the attributes of social, economic or spatial phenomena at one or more discrete moments in time. It is to be clearly distinguished from the alternative mode, which will be discussed later in this chapter and which we may call that of relational dynamics, whose focus is upon the process or processes which transform phenomena between moments in time (contrast this with Wilson's 1981 more formal distinction).

The centrality of this second mode cannot be over-emphasized. Analysis in planning and applied human geography has for years struggled ineffectively with the following fundamental paradox. On the one hand a policy is apparently a normative statement designed to inform the process whereby individuals and groups take deliberate and self-conscious steps to modify the future organization and trends of urban social life. It is active and constructive. On the other, a piece of analysis is apparently no more than a descriptive or exploratory statement whose raw material can be no more than some fraction of the current trends and structures of urban life. It is passive and responsive. Its propositions are inextricably rooted in the trends and structures observed and so incapable of accommodating significant upheaval – the sort of upheaval that major policy initiatives might be expected to achieve. We are thus faced with the uncomfortable possibility that applied human science is irrelevant when the policy questions are non-trivial ones. Even more worrying, however, is the logical extension of this paradox. For it is not unreasonable to characterize a policy not, as above, as a purely normative statement but as one which is itself rooted in some minimal framework of knowledge and understanding. The idea of any purely normative utterance, but none the less one expressed in a natural language, is a patent nonsense. At the very least the terms of which it is constituted must have connotations which are communicable. A policy thus becomes a sort of higher order theoretical statement predicated upon descriptive and explanatory as well as normative propositions. And if this is the case it is but a short step to the conclusion that since policy is in at least this sense unavoidably rooted in analysis its effects are also likely to be rather marginal. The history of planning and urban policy has demonstrated the power – but not the necessity – of this conclusion.

We shall return to the relationship between modes of analysis and policy issues in chapter 8. For the moment it is sufficient to note that such pessimism is hard to avoid if analysis is viewed as no more than is possible within its comparative statics mode. It is to the potential, but more crucially the limitations, of this mode that we must therefore first turn our attention.

Comparative statics

As we have noted the comparative statics mode of analysis is fundamentally about the assembly of a series of snapshots, instantaneous representations of the patterns and relationships of urban social and economic life, into a simple chronological sequence which spans the past and possibly also the imminent future. It therefore facilitates the functions of description and extrapolative forecasting. A description is almost a necessary by-product of any deliberately selective picture of a complex phenomenon. And if, by applying a set of assumptions about a moment of time in the future to that picture, we are left, instantaneously and automatically, with a new and adjusted version of the original, then we have in an obvious way produced a forecast.

All of which is relatively uncontroversial. The problem is that in a rather fundamental sense the dominant comparative statics mode of analysis actually confines us to these two modest levels of investigative achievement. Moreover, the force of this limitation is often not acknowledged in practice – perhaps because of the importance which has for so long been attached to the particular analysis function of forecasting. If future uncertainty can be reduced, then applied human science is generally thought to have fulfilled the most important of its formal requirements. This may be most readily demonstrated by recalling some of the techniques discussed in earlier chapters of this critical review.

In the third chapter, we discussed the approaches most commonly adopted for the broadest levels of analysis in planning and human geography – techniques designed to investigate the overall patterns and structures of the economic and demographic bases of a region, a city or a district. In our discussions we found that the most widely used and comprehensive techniques – namely the input-output and cohort survival models – shared many features in common despite the fundamental differences between the objects of their investigations. In particular, each started off by simply elaborating a definition. In the first case this was the accounting identity that ultimately total receipts must equal total payments, whether for a firm or for a whole economy. And in the second, it was an exhaustive disaggregation of population change, initially into just the three components of natural increase, mortality and migration.

Thus initially at least, the function fulfilled by these techniques is a purely descriptive one. As the definitions are elaborated and their component categories are quantified so we learn something of the local or regional economic or demographic bases. We learn of the dependence of the economy upon certain key manufacturing plants, not just because of the labour forces they employ directly, but also because of the up and down stream linkages in which their current production and trade processes involve them. We learn of varying local rates of infant mortality and patterns of out-migration which fluctuate systematically with the population's age distribution.

Of course the pertinence and practical utility of these descriptions are contingent upon the categories adopted for the elaboration of their respective (definitional) starting points and these classification schemes are not implicit in the initial definitions. Thus it is an important and open theoretical question as to which industrial typology is chosen or whether an age disaggregation is as apposite for migration as it is for survival analysis. We shall return later to theoretical questions. For the moment it is more important to emphasize the independent descriptive function that can be fulfilled by these popular techniques. If relevant and robust typologies are selected (especially in the case of the input-output model) and if relatively modest data collection exercises are feasible (preferably on a regular basis) then the practical value of the methods can be considerable. In each case they focus upon structurally crucial relational mechanisms (transaction and survival). In each case they discipline a practical understanding by ensuring that there are no major gaps in our coverage of economic and demographic bases. And in each case they provide a framework within which to locate less precise and less formal levels of information and understanding.

The fact that this descriptive function is not emphasized at least in the policy-related literature can be explained, once more, simply by reference to the widely felt need amongst planners and policy-makers for predictive analysis. The point was made at the outset of chapter 3 that there exists a basic complementarity between investigative and interventionist functions in planning. When uncertainty about the future cannot be reduced by means of control then it can only be reduced by means of analysis, and such analysis must clearly fulfil a predictive function if it is to be of any use in this respect. Thus we observe most practical interest confined to the predictive use of techniques like input-output analysis, and virtually no planning concern at all with techniques which are only really suited to a descriptive role, like the map pattern analysis models discussed in chapter 4. Moreover, within institutional frameworks which permit little by way of planning control, such as the systems of rules which govern the process of development in the United States, we observe correspondingly greater interest in forecasting analysis than, say, in the United Kingdom which has had for many years a comparatively strong planning system (if only in a negative sense).

Of course within the narrower domain of planning research and analysis itself, interest has also been disproportionately focused upon the predictive functions of techniques. Thus the literature generally emphasizes the forecasting role of the cohort model and the future-orientated applications of input-output analysis, like impact assessment and propulsive planning. Moreover, the technical literature offers a plethora of techniques whose function is more or less exclusively that of forecasting or, more precisely, extrapolation.

Perhaps the best-known, and still most popular, example in planning and human geography is that of the family of spatial interaction models discussed in chapter 4. Essentially statistical devices, they attempt to simplify the aggregate pattern of purpose-specific trip-making at a moment in time and for a given spatial system. In a sense, of course, this formalized simplification, the distance decay function, may be thought of as a description just as a curve indicating the level of total population at a set of dates in the immediate past counts as a simple description. However, the point is that such descriptions are put together in the way that they are primarily with a view to their subsequent use in extrapolation. The inherent interest of the description will normally be sacrificed to the primary goal of a simplified forecast.

The essential methodological difference between these statistical devices which fulfil an extrapolative function and definitional ones which start out at least by fulfilling a more modestly descriptive role lies in the nature of the relationship upon which each is built. As we have seen, definitional devices build elaborate descriptive pictures upon the logical relationship of identity. They simply fragment an aggregate measure by means of a hierarchical process of subdivision. It is a process which is deliberately self-contained. Statistical devices are, by contrast, inductively open. They build upon the relationship of concomitant variation for the purpose of inferring the quantitative properties of one measure from equivalent properties of other ones. The zenith of a piece of statistical analysis is traditionally thought to be the inductive demonstration of a causal relationship. However, in applied research this is typically regarded as no more than an instrumental stage in the quest for a reliable forecast, and the central danger of this extension is that in making it we may lose sight of the intrinsic limits of the basic relationship, namely that of co-variation, upon which it is so precariously founded.

The point is that a statistical technique can do no more than measure concomitant variation. Causal inference must therefore rest upon further non-statistical assumptions. Specifically these are first of all that we accept a purely mechanical or Humean model of causation, and second that within that framework we are able to establish co-variation between dependent and independent measures whilst at the same time demonstrating their relative positions on a chronology and their functional status as a closed causal system. Put bluntly, we must ensure that the cause preceded the effect in real time and that there is no variable excluded from the model which would, if included, require the modification of our causal account.

Even the first of these subordinate conditions is not always easy to satisfy, particularly in spatial analysis. Typically variables will all be measured during one short period of time at a variety of points in a predefined spatial system. To demonstrate subsequently the existence of spatial co-variance is to say nothing about the temporal properties of the relationship.

Moreover this problem palls into insignificance when compared with those raised by the second condition, for this requires a level of structural understanding which is by definition independent of the statistical analysis itself. Thus if that analysis has been pushed to its limits then it will have explored all possible relationships amongst the measured variables. At this point the interpretation of any one relationship in causal terms must depend at the very least upon a non-statistical understanding of the system of interest – one, in other words, in which all variables absent from the analysis are theorized as being structurally irrelevant to the causal nexus.

The attraction of statistical analysis, in policy analysis however, is that it lends itself readily to the purpose of extrapolative forecasting no matter what we may be capable of saying with respect to these conditions. If a relationship of co-variation is statistically demonstrable then this means that it may be expressed as a simple algebraic function. This in turn may be frozen and then extrapolated as far into the future as is consistent both with our willingness to estimate future values of those of its components labelled as independent variables and with our confidence in the temporal stability of its particular functional form.

The superficially remarkable thing is that this practical application of statistical analysis is sometimes proved by the eventual passage of events to be relatively successful. Simple spatial interaction and trend extrapolation models have produced acceptable forecasts, and this without anyone worrying over much about structural theorizing or causal closure. The reason for this is that such models typically measure co-variation between highly aggregated social or economic indicators. The measured spatial or historical co-variation is thus probably best explained in terms of the enormous inertia built into extremely large physical systems and extremely complex social ones. Idiosyncrasies are self-cancelling and thus swamped by the momentum of the overall patterns of physical development and socio-economic evolution.

For the same reasons, the simple definitional devices which lend themselves most readily to descriptive analysis can also be used successfully for forecasting purposes. The mechanisms of change (birth and migration rates, production and trade functions) are frozen on the assumption that their aggregated values represent something of the inertia of numerically large demographic and economic systems. Given such an assumption extrapolative forecasting becomes just as legitimate as with statistical devices, and just as easy. Indeed, no matter how interesting and informative the original description and no matter how clearly this intrinsic interest distinguishes the technique from a crude statistical one, when used for forecasting the two types become indistinguishable. Neither can offer more than the extrapolation of a frozen past.

In other words, despite a multitude of operational and indeed functional differences, each of these techniques and methods reaches its functional limits when used for extrapolative forecasting. Just as statistical techniques offer no more than instantaneous measures of concomitant variation, so descriptive techniques offer no more than instantaneous measures of relational structure (and in the case of demographic techniques, hardly even that). They may all be used for predictive purposes only to the extent, therefore, that we have faith in the aggregate inertia which underpins what we observe as the pattern characterizing one or several unconnected moments in time. It is then in a real sense the framework defined by a comparative statics mode of analysis which constrains a diverse collection of techniques to similar basic functions. As we have noted even the simplest and most mechanical of explanatory or causal accounts depends upon more than the measurement of co-variation. Likewise it depends upon more than the documentation of relational structure or systematic pattern. It requires, in fact, an intellectual process of inductive reasoning which shifts the analyst irrevocably out of a comparative statics mode of analysis.

Relational dynamics

I have chosen to label the alternative mode relational dynamics in order to capture something both of the approach and of the orientation of its overall strategy. The point is to contrast two different research or analytic styles. Whilst the comparative statics mode involves a focus upon selective description at discrete moments in time, the alternative treats directly the processes of transformation and transition. It will be argued below that it is only through the adoption of this less popular mode of analysis that we stand a chance of fulfilling the investigative functions of social explanation and interpretation. Moreover, it is these and only these investigative functions upon which we may subsequently build an approach to planning and urban policy formation capable of achieving appropriate modifications to the underlying trends of urban structures, as we shall argue in chapter 8.

Expressed bluntly, these propositions appear quite unreasonable. It may be true that interpretative analysis inevitably goes beyond that which is possible through aggregate descriptive and statistical quantification. It addresses questions about the motives, purposes and assumptions of the individuals involved in the systems under scrutiny. But to refuse to allow that a concise description or a significant index of statistical association can have a genuinely explanatory function seems nonsensical. After all the idea of statistical or probabilistic explanation is fundamental to the science. Inferential statistics are designed to enable the analyst to fit causal models. Even a description is in an important sense also an explanation. It aims at the broadest level to make the meaning of a complex phenomenon clear. To describe an economy or demographic system is to show didactically how it works, to show how its components are functionally related, to explain it.

None of these sorts of explanation are denied validity in the argument I am advancing in this chapter. The point is that of the various methods of investigation open to the analyst adopting a comparative statics mode, that of inferential statistics probably gets closest to a widely acceptable view of what constitutes social explanation. However, as demonstrated in the preceding section, to infer even a mechanically causal explanation from a variance reducing expression is by definition to do more than simply report that expression. It is to apply intellectual processes – in this case most obviously that of inductive theorizing – to the output of the analysis.

It is obvious, of course, that a variety of intellectual processes are involved throughout any analysis, no matter what its style or mode, and so the dynamic/static distinction is clouded if it is thought to rest only upon the centrality of inductive reasoning. Here we must show that to go beyond description and extrapolation – that is to attempt genuinely social forms of explanation and genuinely human forms of interpretation – is intrinsically to focus upon the points of transition and thus the social processes of transformation. It is thus intrinsically to invoke a distinct mode of analysis. Moreover if this is generally true of all social research and analysis, it must be most true when such analysis is applied to planning and urban policy issues, for in this case the research is required to produce a practical account or explanation. The analysis must in other words be orientated towards action which has as its general goal that of modifying urban institutions and practices. The modification of social structures can only be built upon an analysis of the dynamics of their internal and external relationships. Thus the dependence upon a distinct mode of analysis is reinforced.

Once again the best way of making these points should be by means of examples of the particular techniques that fulfil the requirements of the dynamic mode. Unfortunately, that is where the methods we have reviewed up to this point are seen to be most wanting. In chapter 5 we discussed a series of behavioural research techniques each one of which might have been expected to provide a base upon which a genuinely dynamic approach to location choice analysis could be built. Instead each was found to be myopic with respect both to its theoretical roots and to its manner of application. Typically, studies using each specific method treated a focus upon an immediate object – whether perception, evaluation or action – as theoretically unproblematic. What we have observed in subsequent discussion is that such a simplification is not permissible. There exists an unbreakable circle linking the long-term dynamics of strategic choices – about where the firm's plant should be located or one's own family housed – and the short-term routines of everyday business or family life. Micro level research becomes narrowly descriptive of particular and unique examples of verbal or at least overt behaviour if it ignores this circle. Indeed it often becomes distinguishable from most of the macro level work only in that it describes different objects and these descriptions are couched in more abstruse terminology and are thus frequently more ambiguous.

The real problem is that this state of affairs leaves the analysis of the other side of the circle – that is of the long-term choices themselves – without a micro level framework. To rely upon any of the particular techniques discussed in chapter 5 would be, as we noted at the time, to run the risk of missing the point altogether. Thus studies of structured locational choice have most frequently opted for the easier course of exploring the context of long-term choice (as we saw in chapter 6) rather than the process itself. To treat the process directly would be to raise the broader questions of its micro level basis in experience and attitude formation. To treat it as the product of a context is to remain within the simple safety of a neat and mechanical stimulus-response epistemology. Unfortunately, however, this also means remaining firmly locked into the comparative statics mode of analysis.

The problem that we face, therefore, is to isolate approaches to the practice of planning research which satisfy the requirements of a relational dynamic mode of analysis – and to do so without the benefit of a rich and pertinent investigative tradition upon and from which to build. The only seeds of a genuinely dynamic mode of analysis which we have so far noted were those discussed in the conclusions of chapters 5 and 6, and they appear fundamentally inconsistent.

On the one hand, it would appear that social change may be traced ultimately back to the creative human process of reflexive deliberation. This is the approach to dynamics as a mode of analysis which may be drawn from chapter 5. There the crucial dynamic was said to be the process of long-term or strategic choice and the source of its dynamism, the individual's unique capacity for self-reflection. Thus the focus of research interest suggested by this conclusion is one which is directed towards interpreting the relationship between patterns of experience and the process of attaching meaning to that experience. This becomes the central theoretical relationship because it enables us to trace the sources of social and physical change. Individuals make important long-term choices, which have the effect of modifying their own circumstances and those of other people, by deliberating constructively upon their experiences. And in a complex social world every modification to a part of its tightly woven fabric contributes something to its overall reorganization.

Context impinges upon this approach to analysis in an essentially passive or negative manner. The social, political and economic institutions, like the physical form of the city, constitute a backcloth. At one level they represent the limits of our experience and thus also the raw material of any exercise in reflexive deliberation. Any major choice is therefore firmly grounded in the realities of its context from the outset. At the point of choosing, however, this context may be experienced further as a powerfully constraining medium, one which seriously limits the range of options open at a particular time and thus effectively channels a long-term evolutionary process of personal adaptation. Thus figure 5.4, a crude initial summary of this approach, inevitably located aggregate institutional and physical structure outside the framework of a humanist model of locational choice.

Whilst most of the studies of location choice reviewed in chapter 6 tacitly accepted this passive approach to the analysis of context, they refused to accept the corollary that dynamic analysis could therefore only come from the treatment of the process of choice. However, the study of location choice has, most recently, spawned an alternative approach to dynamic analysis which apparently obviates the need for a detailed micro level interpretative investigation of process. This is the work in the tradition of historical materialism discussed at the end of chapter 6.

If it is accepted that the economic system within which spatial choices are made is itself internally inconsistent and thus intrinsically unstable as a system, then it becomes reasonable to search for the sources of social change in the structures of society themselves. The dynamic key is no longer an individual's self-consciously deliberated intervention to modify the detail of his or her own personal circumstances. It becomes rather the inevitably alienated response to fundamental contradictions built into the circumstances of every member of the society. The focus of research effort suggested by this line of reasoning shifts from the unobservable links between individualized experience and action to the readily observable social relationships of ownership, exploitation and control. Once we acknowledge that these relationships may be endemically conflict-inducing the contextual change becomes intelligible in its own terms and at its own level, not through sympathetic interpretation but through social explanation of a dialectical nature.

How are we to choose between these two, so very different, approaches towards a dynamic mode of analysis? Both offer an account of social and physical change by focusing research effort upon what are characterized as crucial theoretical relationships, but in one case the relationship is locked within the mind of the individual and in the other it is perhaps the most obviously tangible link that exists between the physical and social worlds. The case against a resolution of these two positions has most recently been put (to the detriment of the former) by Kevin Cox in a concluding paper to an updated (1981) version of the seminal collection *Behavioural Problems in Geography*. His argument is that, in moving from a vacuous idealist model which accords some sort of abstractly pure autonomy to the individual, the humanist position necessarily drifts back into an acceptance of a mechanical and thus static view of causation. In acknowledging the explanatory relevance of the external social and physical world, the humanist naturally talks in terms of its constraining or limiting impact upon the self-conscious individual. The analysis thus takes on the static mechanistic shape of traditional positivism.

To any extent that this is true, it is certainly not a tendency confined to humanist analysis. One of the more popular traditions in historical materialism is that of structuralism. In this version, not only does the individual disappear altogether as a relevant subject for study, but so too does dialectics as a central method of analysis. The structure of ownership and control is accorded such a dominating role that individuals are allowed absolutely no independence of their class interests. And the stultifying force of the structural model allows no room for the dialectical analysis of social change.

As many neo-Marxist analysts have argued, this is certainly not a necessary feature of materialist historical analysis. But, by the same token, it is clearly not necessarily the only way in which humanist models may be elaborated or extended. The point may be made by indicating briefly the way in which dialectical explanation at the level of social institutions may complement sympathetic interpretation at the level of the reflective individual. It will not be possible to draw from extensive literatures on either of these approaches because, outside mainstream materialism and conventional behavioural geography (neither of which really satisfy the requirements of the relational dynamics mode), very few relevant studies have been performed.

To give an indication of the complementary roles that may be fulfilled by the two approaches it may be helpful to extend and modify figure 5.4 referred to above so as to accommodate dynamic analysis at the level of the institution as well as at that of the individual or small group. Figure 7.1 contains the original diagram as its inner circle. Thus the long term of assumptions, attitudes and carefully deliberated plans is constantly being fed by a diet of experience, assembled in the course of living out a routinized daily life and mediated by a process of self-aware reflection. The meaning which is in this way attached to experience also informs the so-called 'life choices' we make, and they, in their turn, selectively modify the contexts of our daily lives. We then have to assimilate these contexts adaptively in order to set up routine daily activity patterns. The analyst may gain an understanding of the dynamics of this circle only by means of sympathetic interpretation because the key relationship between objective experience and its subjective meaning is constituted essentially through the process of self-reflection.

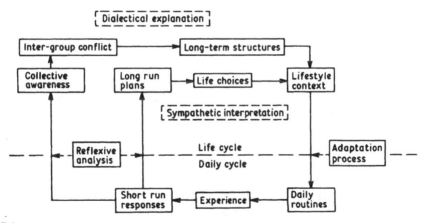

7.1 An integration of dialectical and sympathetic approaches to the dynamics of choice and change

In practice, it is therefore difficult to prescribe operational rules for this approach to relational dynamic analysis. Typically the evidence of an interpretative analysis is only revealed through the ways in which conclusions are drawn in the reports of particular studies. Moreover, most interpretative studies do not link an analysis of the unobservable relationships (the reflexive attachment of meaning to experience) forward to their observable outcomes (the translation of reflexive deliberation into life choices). Typical studies – for instance those addressing the phenomenological concept of place – remain therefore only potentially dynamic (see Relph 1976 for example).

However our own work at University College London has recently attempted to put into practice the strategy outlined above. (The detailed empirical study is reported in Cullen *et al.* 1981.) Briefly we have designed what we term a 'biographical method' for the exploration of the long-term strategic choices of individuals and families. This involves first establishing in some detail (through the use of diary techniques and other retrospective questions) the context of experience against which a recent choice was set. This is used then as the base upon which to build an attempt to recover the meanings imposed upon that experience (through the use of very loosely structured interviewing techniques). And finally, the chronology of the passage through the stages of choice is documented, more or less according to the model of a sequentially unfolding story. The overall method has been used successfully to investigate the process of transition between jobs in an urban context. A somewhat similar approach was adopted by Michelson (1977) in his study of residential choice.

Whatever the interpretative power of such an approach, its ability to yield an account of urban change is logically confined within the limits of our willingness to accept an individualist model of social dynamics as sufficient. Figure 7.1 goes one step further than this in that it also attempts to define a role for relational dynamic investigation at the level of social institutions. In order to complement sympathetic interpretative analysis at the micro level, what is required is a dialectical explanatory account at the level of institutions. The analysis is still grounded ultimately in the same experiential reality, for it can only be through the actions of individuals, even when participating as members of interest groups, that change may occur. However, the extended model acknowledges that the aggregate structure of institutional arrangements may exhibit inconsistencies and inequities which precipitate a different sort of response on the part of those who experience them. Individuals clearly impose very many sorts of meaning upon their circumstances, but if one's research interest is in urban social change, then the critical issue becomes that of the imposition of meanings with a view to action. And in this case, it would seem reasonable to distinguish the processes of reflection which generate an essentially individualized self-awareness or intellectual frame from those which foster what Durkheim (1895) originally termed a collective consciousness. In each case the individual or collective meaning of experience may encompass dissatisfaction or frustration and in each case this may precipitate action. However at the individual or domestic scale, the reflexive image will take the form of a personal plan or family project, designed purely to inform a choice which will have the effect (if effective) of modifying the individual's own circumstances. The fact that these attempted choices will have wider social repercussions does not fundamentally alter the way they must be interpreted. Because of their origins in individualized reflexive thought they will never be clearly intelligible in purely social terms.

Collective awareness, on the other hand, makes sense only if viewed from a broader social perspective. Moreover social science, or more precisely that part of the Hegelian tradition which was subsequently taken over by the materialists, provides us with a method, known as dialectical explanation, for treating the dynamics of social institutions at the social level. The key resides in our ability, through a process of inductive theorizing, to isolate the structural contradictions embedded within a framework of institutions. If such contradictions can be found, and if it may be further theorized that these contradictions have the effect of structuring the alignment of socially coherent groups of the population into relationships of confrontation or conflict, then a genuinely social approach to relational dynamics becomes possible. The dialectical synthesis takes the form of a resolution in one way or another of the conflict which was, and probably still is, implicit in the unstable structure of relationships constituting the institutional system of the time. The system changes over the years as its contradictions and inconsistencies emerge to the collective awareness of relatively homogeneous social groups. They may then become involved as active participants in relationships of confrontation and remould the system in the course of their participation in the process of conflict.

I have chosen to describe the dialectical approach to social explanation in deliberately abstract terms in order to indicate that it is not necessarily a materialist analysis in the traditional sense to which I am referring. Of course the best-known examples are, like those discussed in chapter 6, firmly rooted in that tradition. However there are studies, such as those of Wolpert *et al.* (1972) and Ley and Mercer (1980), which do not adopt a strictly Marxist position and yet do take an essentially dialectical look at the emergence and effects of urban conflicts. The first requirement is simply that the overall theoretical model which is developed should be one which goes beyond the functionalist principles of positivism. Though such principles may help us understand the persistence of social and physical forms, they cannot help us to explain their dynamics since they exclude from consideration the sources of instability and disequilibrium. The theoretical framework must therefore focus upon social relationships within institutional or physical structure which have intrinsically destabilizing effects. Clearly property relationships are included within this category but that does not mean that all others are preempted.

A second requirement of the proposed dialectical approach is that it must be fundamentally historical. The sorts of basic institutional conflict and social instability which constitute the focus of this approach emerge in real time, and generally over a long period. They develop, not just within an historically specific context, but also in response to it. They cannot be understood if abstracted from their place in history and so the narrative methods of historical explanation must be central to any account of the dynamics of urban social institutions.

Conclusions

Kevin Cox was quoted above as espousing this dialectical method whilst dismissing as bourgeois idealism a humanist approach to analysis. He is only the most recent in a long line of materialists who have adopted essentially the same stance. The chief conclusion of our critique so far, however, is that a synthesis can be achieved and that such a synthesis constitutes the only way in which a comprehensively dynamic programme of planning and urban policy analysis may be put together. On what does this optimism rest?

It clearly does not rest upon a large body of research which demonstrates through practice the possibility of effective synthesis. What I am arguing is for the future, and there is no point in pretending that it will be easy. I am, nevertheless, convinced that the effort is worth making for two simple reasons. The first relates to the issue of theoretical or epistemological consistency. Upon this matter, the critical point appears to be that of the shared origins of two eventually very different processes of social change. Whatever the differences between changes that occur through the application of individual self-reflection and those that result from the emergence of collective awareness, they are both rooted ultimately in human social action. Moreover this action is itself based inevitably – given the humanity of its agents – in a process of socially reflexive deliberation. What this commonality means is that to any extent that either level of dynamic analysis renders the other irrelevant or inconsistent, to that extent it must be distorting rather than enhancing our understanding. It must be denying either the self-awareness of its subjects or their membership of human society.

The second reason for persevering in this venture is that so much stands to be gained. The evidence of this book is that depressingly little has so far been achieved through the application of traditional methods and models. This has not been by any means just the fault of the complexity of the subject. The achievement of a practical understanding of urban dynamics is of course never going to be easy, but the evidence of the above review suggests that most of the strategies adopted to date were doomed from the outset because the logic they brought to the task was an inappropriate one. Perhaps by overhauling that logic and accepting that a practical understanding must synthesize individual and social dynamics, the prospects for the future may be improved.

8

Investigation and action

In chapter 7, we attempted to pull together the main threads of the discussion, at least in so far as they bear directly upon the issue of research practice. In one sense this satisfies the terms of reference of the book. The epistemological debate which constituted its trigger has been rendered concrete in two independent senses. It has been focused upon the specific methods and approaches of analysis in planning and human geography in such a way that the limits of their applicability have become capable of more precise definition. And the opposing ends of its spectrum have been brought ·closer together – in a practical if not philosophical sense – by applying their precepts synthetically to the tangible research issue of urban dynamics.

There remains, however, the task of reintegration. If a revitalized research practice can be built securely upon the theoretical base of chapter 7, it is still pertinent to question its relevance within the broader policy domain. The model of policy formation which separates investigation from action was rejected in chapter 2 for reasons that few nowadays would wish to question. However, the implications of breaking down the functional barriers are rarely explored. When they are, this is normally from a materialist perspective (see for instance Scott and Roweis 1977 and Roweis 1981). Within that frame of reference, political or more generally, power relationships are centralized. Policy formation and planning are seen as functions of a local state apparatus which is clearly 'aligned' with respect to these relationships. The unstated goal is the perpetuation or reproduction of these relationships, and analysis loses its autonomy as it becomes purely instrumental to this major goal.

In other words, a strictly materialist reintegration of the processes discussed here moves down a chain of dependence from the relational structure of the society in question through a level of state reinforcement and control to the research practices which inform that level in a purely responsive fashion. The interesting thing about this model is, as we saw in chapter 7, that it is of an intrinsically unstable system because its theoretical apex is a dialectical relationship. This instability may become manifest at any point in time – through a mass movement at the level of social structure to a radical cell at the level of applied research. The model therefore becomes relevant to a study of the relationship between policy formation and research in the context of different social structures and different political philosophies.

In one sense the materialist approach lends itself to this extension for it is an approach to interpreting social and political systems which is intrinsically action-orientated. The problem is, however, that the political and intellectual principles underwriting materialism are inextricable. Thus it exists as a guide to political action purely in relation to a specific social theory which builds upon the concepts of ownership and exploitation. In a mixed economy, the implicit action is that of opposition and the implicit analysis is that of elaborating the social model, with a view to highlighting the system's inconsistencies and so contributing to the implicit action. Little is offered to the policy-maker, planner or applied research worker who is not completely satisfied with the package of social explanation and political intent encapsulated in a materialist thesis. Nor is there much in the way of practical guidance beyond the injunction to oppose. Socialism, as a constructive political philosophy which will operate within a society that is not organized along strictly socialist lines, soon merges imperceptibly with a variety of liberal and social democratic traditions, and the internal coherence of the intellectual and political package is lost.

In effect, the materialist approach may be crudely caricatured as one which captures the indivisibility of the policy-forming and investigative functions of planning, whilst refusing to acknowledge more than one legitimate way of 'managing' that indivisibility. The rest of this chapter will be devoted to a brief examination of an approach other than the revolutionary replacement of one form of indivisibility by another. It should be viewed, not as the last stage of a self-sufficient argument, but rather as a somewhat speculative extension of a partial one. The theory of planning is an exceedingly complex area, and what follows is in no sense a fully worked out alternative theory. It is, however, consistent with what has gone before, and thus, I believe, a relevant if modest elaboration of the overall argument. The first stage of this elaboration must be a reaffirmation of the position reached at the end of chapter 2. There it was proposed that a necessarily integrated approach to planning and urban policy formation might be built, not from the top down, as with materialist models, but from the bottom up. What was intended by this proposition was that any reintegration should focus as much upon the problem of rendering policy formation consistent with analysis as upon the intuitively more reasonable strategy of tailoring analysis to the higher order of political levels of planning and policy action. However, it was never intended that this approach should be implemented in a normative vacuum. Clearly, if a realistic model of the policy forming process is to be developed, it must be one which acknowledges the relativity of everyday practice to some ethical or political principles.

For the moment, however, we shall make no particular assumptions about what those principles are or should be over and above the baseline presupposition that any policy or plan which is being consciously formulated must be intended to have some sort of beneficial impact upon the lives of the members of certain socially or spatially definable groups. In other words nothing is said about specific connotations of the word 'beneficial' nor about the precise social or spatial boundaries around the target groups. Furthermore, nothing is said as to who the policy-makers actually are. Thus the undeniably important issues of the nature and extent of consultation and participation are deliberately ignored. The purpose of accepting these major restrictions is so that we may focus explicit attention upon the rules which govern the process of policy formation, for it is contended here that, quite apart from any political rules which may be imposed, certain logical rules must be obeyed whichever individual or group counts as the policy-maker. It is some of these rules which are discussed below – rules which govern the considerations which become relevant once an attempt is made by anybody to formulate a policy for the purpose of achieving a 'beneficial' impact upon the lives of certain individuals and groups.

It would seem that there are at least four clearly distinguishable rules which should apply in most policy-forming circumstances. These are as follows:

(i)　the policy must, first, specify precisely the groups upon whom beneficial impacts are expected and desired, and, what is much more important, it must specify the units of their life circumstances and/or experience in which such impacts are to be measured;

(ii) it must justify the use of such units in terms of some theoretical statement which relates aspects of life satisfaction for each of the respective groups to variations in the units of their circumstances or experiences measured;

(iii) it must justify the use of such units also in terms of a quite independent theoretical statement which relates the actual implementation of the policy to variations in the measured circumstances or experiences of each of these groups;

(iv) finally, it must set targets or criteria for what are to count as shifts, in each of the measures and for each of the affected groups, which are sufficient to validate the policy as formulated and implemented.

The first and fourth of these conditions are fairly obvious ones, though not as often complied with as one might expect. Together what they involve is the requirement that for any given policy, it must be clear who are the intended beneficiaries, what are the intended effects, and how large must those effects be before the policy can be counted a success. All quite unexceptionable.

Difficulties only really arise with the second and third conditions. This is superficially at least a little surprising since they follow directly from the first. However, what they do, in effect, is to integrate planning and urban policy statements into the overall framework of theoretical social science, and this is what seems semantically inconsistent. For policy statements are traditionally regarded as straight prescriptions, neither true nor false, neither capable of justification by recourse to factual argument nor capable of implying factual conclusions. The instrumental logic of prescriptive statements is assumed to be incapable of absorption into the inductive logic of empirical verification, and vice versa. Yet here we are presenting a set of rules which appear to convert a policy statement into a sort of higher-order theoretical statement which (ideally) contains not only its own criteria for evaluation but also its own logically prior criteria for validation at at least two distinct levels of social theory.

This is not the place to attempt an answer to the age old question in moral philosophy concerning the relationship between 'ought' and 'is'. Nor is it necessary. Those anxious to preserve the logical independence of the ethical/political and descriptive/scientific domains will no doubt be quick to point out that the above argument depends upon at least two ethical or political premises (concerning the meaning of 'beneficial impacts' and the isolation of target groups). Such premises, it will be argued, are themselves incapable of being resolved into a factual debate, and thus remain independent of the apparently technical and instrumental analysis implicit in the four conditions.

It is interesting to note that, in this case at least, the above argument, whilst probably true, still does not guarantee the uniqueness of basic ethical premises. For though it is not possible to verify or reject an ethical or political principle which, for example, selects a particular group for preferential treatment, no more does it seem possible to check the truth or falsity of other premises which appear to be equally crucial to the argument but much closer in form to straightforward descriptive propositions. We shall return to this issue at a later stage in the argument. For the moment the crucial point is, as noted above, that the second and third conditions link the formulation of a policy (no matter what its other connections) directly and necessarily into the framework of theoretical and empirical social science. Moreover, we are not talking about a simple rational sequence of independent and separable steps (some of which are ethical/political and others descriptive/ scientific). We are talking about the logic of the policy statement itself. It cannot even be formulated without begging apparently factual questions. A policy statement must relate directly to a complex observable world of institutions and social practices as well as to a complex ideological one of political values and personal beliefs. It is the relationship between these two levels which constitutes a plan or a policy.

It may be helpful at this point to demonstrate this necessary interdependence by means of a hypothetical example. Suppose that, in accordance with some basic principle and after a little inductive fishing, we come up with the idea of improving the lot of the less mobile members of a city population by the relatively cheap device of defining bus priority lanes. How should we proceed? First we have to define the groups 'at risk' and the experiences we hope to change. We might in this case attempt to pin down the problem by assuming that all those without access to a private car count as 'less mobile', but this still leaves the considerable task of locating and relevantly classifying the subgroups. It leaves also the task of defining the experiences and circumstances which need to be changed. We may decide that the amount of time spent travelling and the spatial and functional variety of activities performed are appropriate 'target' variables. In other words, variation in each of these seems plausibly related to the issue of accessibility which is the area of activity addressed by the policy.

The second and third conditions are designed to test this assumed plausibility by specifying the two relevant theoretical links, first, between the well being of the proposed beneficiaries and the variations in their circumstances and experiences which are intended, and second, between these variations and the mechanisms of the policy itself. Thus in the first case it must be shown that a reduction in travelling time and an increase in the variety of activities performed actually amount to an improvement in the quality of life, however that may be defined. And in the second it must be shown that the provision of bus lanes will actually achieve the specified reductions and increases for the groups concerned. It is immediately clear that a great deal of theoretical and empirical research is presupposed by each of these conditions, even for the simple example chosen. Finally, 'acceptability levels' must be set. These may be specified in units of either objective (the third condition) or subjective (the second condition) experience. Presumably, if the theoretical arguments are both well founded, it is preferable to specify by how much satisfaction levels are to be increased through implementation of the policy, but this may be quite unfeasible. It may only be possible to talk in such crudely quantifiable terms about objective experiences – in the case of this example, travel times and the variety of activities.

This example concerns the day-to-day practice of policy formation and planning. It starts, in other words, after the selection of both a general policy area and a target group. It may thus be argued that the proposed and now exemplified approach still allows a logical wedge to be driven between the fundamentally political or ethical formative stages of policy-making and the fundamentally investigative follow-up stages. In fact, of course, such an interpretation would be inconsistent, not only with the intention of the approach proposed here, but also with the position reached in chapter 2.

The central point – to continue the development of the

theoretical argument initiated earlier in this chapter – is that the second and third rules have the effect of embedding any policy statement into the framework of the social sciences at every level. If, therefore, the argument of this book is acceptable, what follows is that any such statement is contingent at its very highest level upon an appropriate theory of knowledge. Moreover, that theory is in every way and in each of its facets as fundamental and as incapable of empirical verification as is the most basic statement of political or ethical principle.

At this point it will be appropriate to sacrifice the generality of the discussion in order to tie the approach to policy formation back into the specific treatment of applied research developed in chapter 7. In this way it will be possible both to elaborate the model of policy formation in a way which is consistent with the proposed approach to research practice, and to demonstrate the full extent of their interdependence.

The chief constructive conclusion that emerged from chapter 7 was that that which was critical to the enhancement of a practical understanding at the urban scale was a theoretically rigorous approach to the issue of process dynamics. Moreover, the approach found to hold the most potential – both in the sense of its theoretical coherence and in that of its *a priori* pertinence – was one built upon apparently inconsistent intellectual traditions. On the one hand, we argued for an approach that was an essentially humanist one on the grounds that ultimately the source of all urban change rests with the reflexively deliberated choices of individuals. On the other, we allowed that the intrinsically social nature of urban life meant that change might also spring – though mediated through the instrument of human choice – from collective responses to the inconsistent and thus unstable nature of social and institutional structure.

If it is legitimate to base an approach to analysis in planning and human geography upon these intellectual principles then it must also be inconsistent to deny policy statements and plans a similar intellectual basis. At the earliest and most loosely constrained stages of policy formation, the espousal of such principles will inevitably have an effect. If, first, we adopt a dialectical approach to the practice of an applied human science, this means that we accept that the organization of society is internally inconsistent in some way or ways to an extent sufficient to generate destabilizing conflict. Certain groups are structurally interrelated such that one may plausibly be thought of – and may come to think of itself – as inequitably or unjustly related to another. The interpretation contains its own political ethic, which becomes in a sense the motor of social change. In this intellectual context it is inconceivable that the ethical or ideological basis of policy formation could be insulated from its descriptive basis because the latter is rooted in an interpretation which is intrinsically value laden.

Of course exactly the same is true if we adopt a humanist approach to analysis. In this case we focus in the first instance upon the individual rather than the social context, and to this individual we grant a certain reflexive autonomy. Within this intellectual frame the concepts of belief, attitude and value all become relevant and problematic. They cease to be arbitrary labels for observable phenomena when they become parts of the process whereby meaning is self-consciously ascribed to the observable features of everyday life. But as they acquire such a pivotal intellectual status so also do they help to form the ethical and political framework in which policies are constructed. They demand an approach to policy formation which is intrinsically sympathetic to the aspirations and frames of reference of the individuals who constitute the client populations. Again the idea of insulating the ethical level from the descriptive level when the latter is based upon a premise as fundamental as a belief in the ultimate independence of the human will is shown to be a patent nonsense.

The point of this extended discussion of the logic of the policy-making process is not just to indicate the complexity of the whole thing, for much of what is actually involved has been deliberately ignored. Indeed, the whole process has been approached largely from the point of view of the planning analyst. The point worth taking from the discussion is not that policy-making and research in planning are very difficult, but that they are essentially one and the same task. If the rules discussed and exemplified above are accepted as something like part of what policy-making should be, then, without losing anything of its political, ethical or prescriptive nature, a policy statement becomes also a higher-order theoretical statement which begs as many substantive questions as it does normative ones. And it is only if these substantive questions are specified that applied research can be usefully attempted, for a complete answer to each will amount to a measurement of the policy's coherence, relevance and impact. If the policy is the theory, then its analysis becomes the test of that theory. Planning and urban policy making are thereby tied firmly to an understanding of the social world in which they must succeed, and applied research is at last lent an immediacy and relevance for which it has so long struggled without conspicuous success.

Furthermore, if the substantive levels of a policy statement are built upon the approach to analysis which we have labelled relational dynamics, then it also becomes possible to think in terms of a practical knowledge base which is itself creative or active. The central argument of chapter 7 was that the two most viable routes to dynamic analysis outlined above can be rendered consistent only if we accept an essentially humanist intellectual starting point. This means granting a degree of reflexive autonomy to the individual. If this argument is accepted then the emergence of the collective awareness of an inconsistency in social structure and the formulation of a long-term personal plan of action each become explicable in terms of the same creative reflection upon day-to-day experience. From the policy perspective, however, the crucial point is that whichever way we choose to build further theories upon this reflexive intellectual base, we end up with an interpretation or understanding which contains the rudiments of a humanized model of urban social change. As investigation is focused upon the two crucial levels of urban dynamics, so policy is afforded the basic means for breaking out of the moulds set by the past and the present. And as both investigation and policy are tied to the same humanist epistemology, the moulds should get broken in ways which remain sensitive to the shared values of the societies which have chosen to adopt this approach to applied urban analysis.

References

Ackoff, R.L. (1974) *Redesigning the Future: A Systems Approach to Societal Problems*, New York, Wiley.

Ackoff, R.L. and Sasieni, M.W. (1968) *Fundamentals of Operations Research*, New York, Wiley.

Alonso, W. (1964) *Location and Land Use*, Cambridge, Mass., Harvard University Press.

Althusser, L. (1969) *For Marx*, Harmondsworth, Penguin.

Archer, B.H. (1976) 'The anatomy of a multiplier', *Regional Studies*, 10, 71-7.

Arrow, K.J. (1951) *Social Choice and Individual Values*, New York, Wiley.

Ashby, W.R. (1956) *An Introduction to Cybernetics*, London, Chapman & Hall.

Ashcroft, B. and Swales, J.K. (1982) 'The importance of the first round in the multiplier process', *Environment and Planning A*, 14, 429-44.

Atkins, R.H. (1974) *Mathematical Structure in Human Affairs*, London, Heinemann.

Ball, M.J. (1973) 'Recent empirical work on the determinants of relative house prices', *Urban Studies*, 10, 213-33.

Ball, M.J. (1978) 'British housing policy and the house building industry', *Capital and Class*, 4, 78-99.

Ball, M.J. (1979) 'A critique of urban economics', *Journal of Urban and Regional Research*, 3, 309-32.

Ball, M.J. and Kirwan, R.M. (1977) 'Accessibility and supply constraints in the urban housing market', *Urban Studies*, 14, 11-32.

Bannister, D. and Mair, J.M.M. (1968) *The Evaluation of Personal Constructs*, New York, Academic Press.

Barras, R. and Broadbent, T.A. (1979) 'The analysis in English structure plans', *Urban Studies*, 16, 1-18.

Barras, R. and Broadbent, T.A. (1981) 'A review of operational methods in structure planning', *Progress in Planning*, 17, 53-123.

Barrett, F.A. (1976) 'The search process in residential relocation', *Environment and Behaviour*, 8, 169-98.

Bassett, K. and Short, J.R. (1980) *Housing and Residential Structure*, London, Routledge & Kegan Paul.

Batty, M. (1976) *Urban Modelling: Algorithms, Calibrations, Predictions*, Cambridge, Cambridge University Press.

BBC (1975) *The People's Activities and the Use of Time*, London, BBC Publications.

Beckmann, M. and Marshak, T. (1955) 'An activity analysis approach to location theory', *Kyklos*, 8, 125-43.

Bennett, R.J. (1975) 'The representation and identification of spatio-temporal systems: an example of population diffusion in north-west England', *Transactions of the Institute of British Geographers*, 66, 73-94.

Bennett, R.J. (1976) 'Non-stationary parameter estimation for a small sample situation: a comparison of methods', *International Journal of Systems Science*, 7, 257-75.

Bentham, J. (1789) *The Principles of Morals and Legislation*, London.

Berger, P.L. and Luckmann, T. (1967) *The Social Construction of Reality*, London, Allen Lane.

Bertalanffy, L. (1972) *General Systems Theory*, London, Allen Lane.

Bhaskar, R. (1979) *The Possibility of Naturalism*, Brighton, Harvester Press.

Billings, R.B. (1969) 'The mathematical identity of the multipliers derived from the export base model and the input-output model', *Journal of Regional Science*, 9, 471-3.

Boddy, M.J. (1976) 'The structure of mortgage finance', *Transactions of the Institute of British Geographers*, 1, 20-33.

Bourdieu, P. (1979) *Outline of a Theory of Practice*, Cambridge, Cambridge University Press.

Bracken, I. (1982) 'New directions in key activity forecasting', *Town Planning Review*, 53, 51-64.

Bracken, I. and Hume, D. (1981) 'Forecasting methods and techniques in structure plans', *Town Planning Review* 52, 375-89.

Breheny, M.J. and Roberts, A.J. (1978) 'An integrated forecasting system for structure planning', *Town Planning Review* 49, 306-18.

Broadbent, T.A. (1978) *Planning and Profit in an Urban Economy*, London, Methuen.

Brown, D.J. (1979) 'The location decision of the firm', *Regional Science Association Papers* 43, 23-39.

Brown, H.J. (1969) 'Shift and share projections of regional economic growth', *Journal of Regional Science* 9, 1-18.

Brown, L.A. and Moore, E.G. (1970) 'The intra-urban migration process: a perspective', *Geografiska Annaler* 52b, 1-13.

Brownrigg, M. (1973) 'The economic impact of an urban university', *Scottish Journal of Political Economy* 20, 123-39.

Brownrigg, M. (1980) 'Industrial contraction and the regional multiplier effect', *Town Planning Review*, 51, 195-210.

Buttimer, A. (1974) 'Values in geography', *Association of American Geographers, Commission on College Geography, Resource Paper* 24, Washington, DC.

Buttimer, A. (1976) 'Grasping the dynamism of the life-world', *Annals of the Association of American Geographers* 66, 277-92.

Buttler, H.-J. (1981) 'Equilibrium of a residential city, attributes of housing and land use zoning', *Urban Studies* 18, 23-39.

Camhis, M. (1979) *Planning Theory and Philosophy*, London, Tavistock.

Cant, D.H. (1976) 'Squatting and private property rights', *Town Planning Discussion Paper*, 24, Bartlett School, University College London.

Carrothers, G.A.P. (1956) 'An historical review of the gravity and potential concepts of human interaction', *Journal of the American Institute of Planners* 22, 226-42.

Castells, M. (1972) *La Question Urbaine*, Paris, Maspero; translated 1977 as *The Urban Question*, London, Edward Arnold.

Castells, M. (1976) 'Theoretical propositions for an experimental study of urban social movements', in Pickvance, C.G. (ed.) *Urban Sociology*, London, Tavistock.

Catanese, A.J. and Steiss, A.W. (1970) *Systems Planning: Theory and Application*, Lexington, Mass, Lexington Books.

Chadwick, G. (1971) *A Systems View of Planning*, Oxford, Pergamon.

Chalmers, J. A. and Beckhelm, T. L. (1976) 'Shift and share and the theory of industrial location', *Regional Studies* 10, 15-23.

Chapin, F.S. (1974) *Human Activity Patterns in the City*, New York, Wiley.

Chisholm, M. (1971) 'In search of a basis for location theory', *Progress in Geography* 3, 111-33.

Cleveland County Planning Team (1975) *The Economic Impact of North Sea Oil in Cleveland*, Middlesbrough, Cleveland County Council.

Cockburn, C. (1977) *The Local State*, London, Pluto Press.

Cole, R. L. (1974) *Citizen Participation and the Policy Process*, Lexington, Mass., D. C. Heath.

Coupe, R.T. and Morgan, B.S (1981) 'Towards a fuller understanding of residential mobility', *Environment and Planning A* 13, 201-15.

Cox, K.R. (1981) 'Bourgeois thought and the behavioural geography debate', in Cox, K.R. and College, R.G. (eds) *Behavioural Problems in Geography Revisited*, London, Methuen.

Cullen, I.G. (1976) 'Human geography, regional science, and the study of individual behaviour', *Environment and Planning A* 8, 397-409.

Cullen, I.G. (1978) 'The treatment of time in the explanation of spatial behaviour' in Carlstein, T., Parkes, D.N. and Thrift, N.J. (eds) *Human Activity and Time Geography*, London, Edward Arnold.

Cullen, I.G. (1983) 'Expert systems in architectural and planning education', *Proceedings of 2nd International ECAADE Conference*, Brussels, Free University.

Cullen, I.G. (1984) 'Q-analysis and the theory of social scientific knowledge', *Environment and Planning B* (forthcoming).

Cullen, I.G. and Godson, V.(1975) 'The structure of activity patterns', *Progress in Planning* 4, 1-96, Oxford, Pergamon.

Cullen, I.G. and Phelps, E. (1975) *Diary Techniques and the Problems of Urban Life*, Final Report to SSRC (HR2336).

Cullen, I.G., Hammond, S. and Haimes, E.V.(1980) *Travel and Urban Working Life Styles 1974-1978: A Longitudinal Diary Study in an Inner London Council Estate*, Final Report to the Transport and Road Research Laboratory (TRR 842/2 78Z).

Cullen, I.G., Hammond, S. and Haimes, E.V. (1981) *Employment and Mobility in Inner Urban Areas: An Interpretive Study*, Final Report to SSRC (HR 5884), Bartlett School of Architecture and Planning.

Curry, L. (1971) 'Applicability of space-time moving average forecasting', in Chisholm, M. *et al.* (eds) *Regional Forecasting*, London, Butterworth.

Curry, L. (1972) 'A spatial analysis of gravity flows', *Regional Studies* 6, 131-47.

De Kanter, J. and Morrison, W. (1978) 'The Merseyside input-output study and its application to structure planning', in Batey, P. W. J. (ed.) *Theory and Methods in Urban and Regional Analysis*, London, Pion.

Dear, M. and Scott, A.J. (1981) *Urbanisation and Urban Planning in Capitalist Society*, London, Methuen.

Department of the Environment (1977) *Policy for the Inner Cities*, London, HMSO.

Diamond, D.B. (1980) 'Income and residential location: Muth revisited', *Urban Studies*, 17, 1-12.

Dicken, P. (1976) 'The multi-plant business enterprise and geographic space', *Regional Studies* 10, 401-12.

Dickens, P. (1977) 'Social change, housing and the state: some aspects of class fragmentation and incorporation', in Harloe, M. (ed.) *Second Conference on Urban Change and Conflict*, CES Conference Papers, London.

Downs, R. and Stea, D. (1977) *Maps in Minds*, New York, Harper & Row.

Dror, Y. (1968) *Public Policy Making Re-examined*, Scranton, Pa., Chandler.

Durkheim, E. (1895) *Les Règles de la Méthode Sociologique*, Paris, Felix Alaan; translated 1964 as *Rules of Sociological Method*, New York, Free Press.

Erickson, R.A., (1980) 'Corporate organisation and manufacturing branch plant closures in nonmetropolitan areas', *Regional Studies* 14, 491-501.

Evans, A. (1973) *The Economics of Residential Location*, London, Macmillan.

Eversley, D. (1973) *The Planner in Society*, London, Faber & Faber.

Fagence, M. (1977) *Citizen Participation in Planning*, Oxford, Pergamon.

Feigenbaum, E.A. and Barr, A. (1982) *The Handbook of Artificial Intelligence*, London, Pitman (three volumes).

Filmer, P., Phillipson, M., Silverman, D. and Walsh, D. (1972) *New Directions in Sociological Theory*, London, Collier-Macmillan.

Foot, D. (1981) *Operational Urban Models*, London, Methuen.

Ford, R.G. and Smith, G.C. (1981) 'Spatial aspects of inter-urban migration behaviour in a mixed housing market', *Environment and Planning A* 13, 355-71.

Fothergill, S. and Gudgin, G. (1982) *Unequal Growth: Urban and Regional Employment Change in the UK*, London, Heinemann.

Friend, J.K. (1980) 'Planning in a multi-organisation context', *Town Planning Review* 51, 261-9.

Friend, J.K. and Jessop, W.N. (1969) *Local Government and Strategic Choice*, London, Tavistock.

Friend, J.K., Power, J.M. and Yewlett, C.J.L. (1974) *Public Planning: The Inter-Corporate Dimension*, London, Tavistock.

Galbraith, J.K. (1967) *The New Industrial State* (2nd edition) Boston, Houghton Mifflin.

Garnick, D.H. (1970) 'Differential regional multiplier models', *Journal of Regional Science* 10, 35-48.

Gatrell, A.C. (1981) 'On the structure of urban social areas: explorations using Q-analysis', *Transactions of the Institute of British Geographers* n.s. 6, 228-45.

Gershuny, J.I. and Thomas, G.S. (1981) 'Changing patterns of time use', *Science Policy Research Unit Occasional Paper* 13, Brighton, SPRU.

Giddens, A. (1979) *Central Problems in Social Theory*, London, Macmillan.

Giddens, A. (1981) *A Contemporary Critique of Historical Materialism*, London, Macmillan.

Gilmour, J. (1974) 'External economies of scale, interindustrial linkages, and decision making in manufacturing', in Hamilton, F. (ed.) *Spatial Perspectives on*

Industrial Organisation and Decision Making, New York, Wiley.

Goddard, J.B. (1973) 'Office linkages and location, progress in planning 1, 109-232, Oxford, Pergamon.

Goddard, J.B. and Morris, D.M. (1976) 'The communications factor in office decentralization', *Progress in Planning* 6, 1-80, Oxford, Pergamon.

Goldman, T. (1958) 'Efficient transportation and industrial location', *Regional Science Association Papers* 4, 103-24.

Goldner, W. (1971) 'The Lowry model heritage', *Journal of the American Institute of Planners* 37, 100-10.

Goldsmith, M. and Saunders, P. (1975) 'The tale of Lewis and the cat: public participation and the settlement policy in Cheshire', *Linked Research Project, Sheffield Research Paper 8*.

Gould, P. (1980) 'Q-analysis, or a language of structure: an introduction for social scientists, geographers and planners', *International Journal of Man-Machine Studies*, 12, 169-99.

Gould, P. (1981) 'Letting the data speak for themselves', *Annals of the Association of American Geographers* 71, 166-75.

Greater Manchester Planning Team (1975) *County Structure Plan: Report of Survey: Employment and the Economy*, Greater Manchester Council.

Gregory, D. (1978) *Ideology, Science and Human Geography*, London, Hutchinson.

Gregory, D. (1981) 'Human agency and human geography', *Transactions of the Institute of British Geographers* 6, 1-18.

Greig, M.A. (1971) 'The regional income and employment effects of a pulp and paper mill', *Scottish Journal of Political Economy* 18, 31-48.

Habermas, J. (1972) *Knowledge and Human Interests*, London, Heinemann.

Hagerstrand, T. (1970) 'What about people in regional science?', *Papers of the Regional Science Association* 24, 7-21.

Hagerstrand, T. (1974) 'Time-geographic notation: purposes and premises', *Swedish Geographical Yearbook* 50, 86-94.

Haggett, P. *et al.* (1977) *Locational Analysis in Human Geography: Locational Methods* (2nd edition), London, Edward Arnold.

Hall, P. (1981) *The Inner City in Context*, London, Heinemann.

Harloe, M. (ed.) (1977) *Captive Cities*, Chichester, Wiley.

Harris, C.D. and Ullman, E.L. (1945) 'The nature of cities' *Annals of the American Academy of Political and Social Sciences*, 242, 7-17.

Harvey, D. (1969) *Explanation in Geography*, London, Edward Arnold.

Harvey, D. (1973) *Social Justice and the City*, London, Edward Arnold.

Harvey, D. (1977) 'Labour, capital and class struggle around the built environment in advanced capitalist societies', *Politics and Society* 6, 265-95.

Harvey, D. (1981) *The Limits to Capital*, Oxford, Blackwell.

Herbert, D.T. (1973) 'The residential mobility process', *Area* 5, 44-8.

Herbert, D.T. and Johnston, R.J. (1976) *Social Areas in Cities*, London, Wiley (two volumes).

Herbert, D.T. and Smith, D.M. (1979) *Social Problems and the City: Geographical Perspectives*, Oxford, Oxford University Press.

Herbert, J.D. and Stevens, B.H. (1960) 'A model of the distribution of residential activity in urban areas', *Journal of Regional Science* 2, 21-36.

Hirschman, A.O. (1958) *The Strategy of Economic Development*, New Haven, Conn., Yale University Press.

Hoinville, G. (1971) 'Measuring environmental preferences', *Environment and Planning A* 3, 33-50.

Holtermann, S. (1975) 'Areas of urban deprivation in Great Britain: an analysis of 1971 census data', *Social Trends* 6, 37-47.

Hordijk, L. and. Nijkamp, P. (1977) 'Dynamic models of spatial auto-correlation', *Environment and Planning A* 9, 37-47.

Hotelling, H. (1929) 'Stability in competition', *Economic Journal* 39, 31-57.

Houston, D. (1967) 'Shift-share analysis: a critique', *Southern Economic Journal* 30, 557-81.

Hoyt, H. (1939) *The Structure and Growth of Residential Neighborhoods in American Cities*, Washington, DC, US Government Press Office.

Ingram, G.K. *et al.* (1972) *The Detroit Prototype of the NBER Urban Simulation Model*, New York, Columbia University Press.

Isard, W. *et al.* (1960) *Methods of Regional Analysis*, Cambridge, Mass., MIT Press.

Isard, W. and Smith, T.E. (1967) 'Location games with applications to classic location problems', *Regional Science Association Papers* 19, 45-80.

Jencks, C. (1972) *Inequality: A Reassessment of the Effects of Family and School in America*, New York, Basic Books.

Johnson, K.H. and Lyon, H.L. (1973) 'Experimental evidence on combining cross-section and time series information', *Review of Economics and Statistics* 55, 465-74.

Jones, P.M. (1979) 'HATS: a technique for investigating household decisions', *Environment and Planning A* 11, 59-70.

Kain, J.F. and Quigley, J.M. (1975) *Housing Markets and Racial Discrimination*, New York, NBER.

Keeble, D. (1976) *Industrial Location and Planning in the UK*, London, Methuen.

King, L.J. (1969) *Statistical Analysis in Geography*, Englewood Cliffs, N.J. Prentice-Hall.

Koopmans, T.C. and Beckmann, M.J. (1957) 'Assignment problems and the location of economic activities', *Econometrica* 25, 53-76.

Kuhn, H.W. and Kuenne, R.E. (1962) 'An efficient algorithm for the solution of the generalized Weber problem in space economies', *Journal of Regional Science* 4, 21-33.

Kuhn, T. (1970) *The Structure of Scientific Revolutions*, Chicago, University of Chicago Press.

Launhardt, W. (1885) *Mathematische Begründung der Volkwirtschaftslehre*, Leipzig.

Layder, D. (1981) *Structure, Interaction and Social Theory*, London, Routledge & Kegan Paul.

Lee, T. (1970) 'Perceived distance as a function of direction in the city', *Environment and Behaviour* 2, 40-51.

Lenntorp, B. (1976) 'Paths in space-time environments', ·*Lund Studies in Geography*, B, 44, Lund.

Leontieff, W. (1966) *Input-Output Economics*, Oxford, Oxford University Press.

Ley, D. and Mercer, J. (1980) 'Locational conflict and the politics of consumption', *Economic Geography* 56, 89-109.

Lichfield, N., Kettle, P. and Whitbread, M. (1975) *Evaluation in the Planning Process*, Oxford, Pergamon.

Little, I.M.D. (1952) 'Social choice and individual values', *Journal of Political Economy* 60, 422-32.

Losch, A. (1940) *The Economics of Location*, New Haven, Conn., Yale University

Press; 1954 translation by W. H. Woglom.

Lowry, I. S. (1964) *A Model of Metropolis*, Santa Monica, Cal., Rand (RM-4035-RC).

MacKay, D.I. (1968) 'Industrial structure and regional growth: a methodological problem', *Scottish Journal of Political Economy* 16, 129-43.

McLoughlin, J.B. (1969) *Urban and Regional Planning: A Systems Approach*, London, Faber & Faber.

McNicholl, I.H. and Walker, G. (1979) 'The Shetland economy 1976/7: structures and performance', *Shetland Times* (Lerwick).

Martin, R.L. and Oeppen, J.E. (1975) 'The identification of regional forecasting models using space-time correlation functions', *Transactions of the Institute of British Geographers* 66, 186-97.

Marx, K. (1867) *Capital*, published 1976, Harmondsworth, Penguin.

Masser, I. and Brown, P. (1977) 'Spatial representation and spatial interaction', *Papers of the Regional Science Association* 38, 71-92.

Massey, D.B. (1974) 'Towards a critique of industrial location theory', *Centre for Environemntal Studies*, Research Paper 5.

Massey, D.B. (1978) 'Capital and locational change', *Review of Radical Political Economics* 10, 39-54.

Massey, D.B. and Meegan, R. (1978) 'Industrial restructuring versus the cities', *Urban Studies* 15, 273-88.

Massey, D.B. and Meegan, R. (1979) 'The geography of industrial reorganisation', *Progress in Planning* 10, 155-237.

Massey, D.B. and Meegan, R. (1982) *The Anatomy of Job Loss*, London, Methuen.

Merrett, A.J. and Sykes, A. (1973) *The Finance and Analysis of Capital Projects*, London, Longman.

Michelson, W. (1977) *Environmental Choice, Human Behaviour and Residential Satisfaction*, New York, Oxford University Press.

Miernyk, W.H. (1970) *Simulating Regional Economic Development: An Interindustry Analysis of the West Virginia Economy*, Lexington, D. C. Heath.

Milgram, S. (1972) 'The experience of living in cities', in Mesarovic, M. and Reisman, A. (eds) *Systems Approach and the City*, Amsterdam, North Holland.

Mill, J.S. (1867), *Utilitarianism*, (3rd edition), London.

Mishan, A.J. (1969) *Welfare Economics: An Assessment*, Amsterdam, North Holland.

Moser, C. A. and Scott, W. (1961) *British Towns*, Edinburgh, Oliver & Boyd.

Moses, L. (1955) 'The stability of interregional trading patterns and input-output analysis', *American Economic Review* 45, 803-32.

Muth, R.F. (1969) *Cities and Housing*, Chicago, University of Chicago Press.

Myrdal, G. (1957) *Economic Theory and Underdeveloped Regions*, London, Duckworth.

Northern Region Strategy Team (1976) *Growth and Structural Change in the Economy of the Northern Region since 1952*, Technical Report 4, Newcastle, NRST.

Ohls, J.C. *et al.* (1974) 'The effects of zoning and land value', *Journal of Urban Economics* 1, 428-44.

Olsson, G. (1965) *Distance and Human Interaction*, Philadelphia, Regional Science Research Institute.

O'Neill, J. (1973) *Modes of Individualism and Collectivism*, London, Heinemann.

Openshaw, S. (1978) 'An empiricial study of some zone design criteria', *Environment and Planning A* 10, 781-94.
Park, R.E., Burgess, E.W. and McKenzie, R.D. (1925) *The City*, Chicago, University of Chicago Press.
Parkes, D.N. and Thrift, N.J. (1980) *Times, Spaces and Places: A Chronogeographic Perspective*, Chichester, Wiley.
Perraton, J. and Baxter, R. (eds) (1974) *Models, Evaluations and Information Systems for Planners*, Lancaster, MTP Construction Press.
Popper, K. (1961) *The Poverty of Historicism*, London, Routledge & Kegan Paul.
Popper, K. (1976) 'The logic of the social sciences', in Adorno, T. *et al. The Positivist Dispute in German Sociology*, London, Heinemann.
Poulantzas, N. (1975) *Political Power and Social Classes*, London, New Left Books.
Pred, A. (1967) 'Behaviour and location', *Lund Studies in Geography*, B,1,2, Lund.
Pred, A. (1981) 'Social reproduction and the time geography of everyday life', *Geografisker Annaler*, 63 (B), 5-22.
Pressat, R. (1978) *Statistical Demography*, London, Methuen.
Putman, S.H. (1980) 'Calibrating urban residential location models 3', *Environment and Planning A*, 12, 813-27.
Randall, J.N. (1973) 'Shift-share as a guide to the employment performance of west central Scotland', *Scottish Journal of Political Economy* 20, 1-26.
Rawls, J. (1972) *A Theory of Justice*, Oxford, Clarendon Press.
Rees, J. (1979) 'Technological change and regional shifts in American manufacturing', *Professional Geographer* 31, 45-55.
Rees, P.H. and Wilson, A.G. (1977) *Spatial Population Analysis*, London, Edward Arnold.
Rein, M. (1976) *Social Science and Public Policy*, Harmondsworth, Penguin.
Relph, E. (1976) *Place and Placelessness*, London, Pion.
Rhodes, J. and Moore, B. (1973) 'Understanding the effects of British regional policy', *Economic Journal* 83, 87-100.
Richardson, H.W. (1977) *The New Urban Economics*, London, Pion.
Richardson, H.W. *et al.* (1975) *Housing and Urban Spatial Structure*, Farnborough, Hants, Saxon House.
Riefler, R. and Tiebout, C.M. (1970) 'Interregional input-output: an empirical California-Washington model', *Journal of Regional Science* 10, 135-52.
Robinson, J.P. (1977) *Changes in American Use of Time: 1965 – 1975*, Cleveland, Ohio, Communications Research Centre, Cleveland State University.
Robinson, J.P. (1978) 'Time use as a social indicator', in Michelson, W. (ed.) *Public Policy in Temporal Perspective*, The Hague, Mouton.
Rosenhead, J. and Gupta, S.K. (1968) 'Robustness in sequential investment decisions', *Management Science* 15, 8-18.
Rossi, P. (1955) *Why Families Move*, Glencoe, Ill., Free Press.
Roweis, S.T. (1981) 'Urban planning in early and late capitalist societies', in Dear, M. and Scott, A.J. (eds) *Urbanization and Urban Planning in Capitalist Society*, London, Methuen.
Roweis, S.T. amd Scott. A.J. (1978) 'The urban land question', in Cox, W. (ed.) *Urbanization and Conflict in Market Societies*, Chicago, Maaroufa.
Rowley, G. and Tipple, G. (1974) 'Coloured immigrants within the city: an analysis of housing and travel preferences', *Urban Studies* 11, 81-9.
Rowley, G. and Wilson, S. (1975) 'The analysis of housing and travel preferences: a gaming approach', *Environment and Planning A*, 7, 171-7.

Salling, M. and Harvey, M.E. (1981) 'Poverty, personality and sensitivity to residential stressors', *Environment and Behaviour* 13, 131-63.

Samuelson, P. A. (1939) 'Interactions between multiplier analysis and the principle of acceleration', *Review of Economics and Statistics*, 21, 75-8.

Samuelson, P.A. (1947) *Foundations of Economic Analysis*, Cambridge, Mass., Harvard University Press.

Samuelson, P.A. (1967) 'Arrow's mathematical politics', in Hook, S. (ed.) *Human Values and Economic Policy*, New York, New York University Press.

Sayer, R.A. (1976) 'A critique of urban modelling', *Progress in Planning* 6, 187-254, Oxford, Pergamon.

Sayer, R.A. (1978) 'Mathematical modelling in regional science and political economy: some comments', *Antipode* 10, 79-86.

Sayer, R.A. (1979) 'Understanding urban models versus understanding cities', *Environment and Planning* A 11, 853-62.

Scott, A.J. and Roweis, S.T. (1977) 'Urban planning in theory and practice: a reappraisal', *Environment and Planning A* 9, 1097-119.

Sen, A.K. (1970) *Collective Choice and Social Welfare*, San Francisco, Holden Day.

Senior, M.L. and Wilson, A.G. (1974) 'Explorations and syntheses of linear programming and spatial interaction models of residential location', *Geographical Analysis* 6, 209-38.

Shevky, E. and Bell, W. (1955) *Social Area Analysis*, Stamford, Conn.

Short, J.H. (1978) 'Residential mobility', *Progress in Human Geography* 2, 419-48.

Simon, H.A. (1959) *Models of Man*, New York, Wiley.

Smith, D.M. (1966) 'A theoretical framework for geographical studies of industrial location', *Economic Geography* 42, 95-113.

Smith, D.M. (1971) *Industrial Location*, New York, Wiley.

Smith, D.M. (1975) *Patterns in Human Geography*, Harmondsworth, Penguin.

Speare, A. *et al.* (1974) *Residential Mobility, Migration and Metropolitan Change*, Cambridge, Mass., Ballinger.

Stea, D. (1969) 'The measurement of mental maps' in Cox, K.R. and Golledge, R.G. (eds) *Behavioural Problems in Geography*, Evanston, Ill., Northwestern University Press.

Stegman, M.A.(1969) 'Accessibility models and residential location', *Journal of the American Institute of Planners* 35, 22-9.

Stevens, B.H. (1961) 'An application of game theory to a problem in location strategy', *Regional Science Association Papers* 7, 143-57.

Storper, M. (1981) 'Toward a structural theory of industrial location', in Rees, J. *et al.* (eds) *Industrial Location and Regional Systems*, London, Croom Helm.

Stringer, P. and Ewens, S. (1974) 'Participation through public meetings: the case in N.E. Lancashire', *Linked Research Project, Sheffield, Research Paper* 2.

Szalai, A. (1972) *The Use of Time*, The Hauge, Mouton.

Tabb, W.K. and Sawers, L. (1978) *Marxism and the Metropolis*, Oxford, Oxford University Press.

Thorngren, B. (1970) 'How do contact systems affect regional development', *Environment and Planning* 2, 409-27.

Thornley, A. (1977) 'Theoretical perspectives on planning participation', *Progress in Planning* 7, 1-157, Oxford, Pergamon.

Thrift, N. (1983) 'On the determination of social action in space and time', *Environment and Planning D: Society and Space* 1, 23-57.

Thunen, J.H. von (1826) *Der Isolierte Staat*, Hamburg; translated 1966 by C.M Wartenburg, Oxford, Oxford University Press.

Tiebout, C.M. (1962) *The Community Economic Base Study, Committee for Economic Development*, Supplementary Paper No. 16.

Tobler, W.R.(1973) 'Regional analysis: time series extended to two dimensions', *Geographia Polonica* 25, 101-6.

Tornquist, G. (1977) 'The geography of economic activities', *Economic Geography* 53, 153-62.

Townroe, P. (1972) 'Some behavioural considerations in the industrial location decision', *Regional Studies* 6, 261-72.

Townroe, P. (1975) 'Branch plants and regional development', *Town Planning Review* 46, 47-62.

Townroe, P. (1976) *Planning Industrial Location*, London, Leonard Hill.

Webber, R.J. (1978) 'Making the most of the canons for strategic analysis', *Town Planning Review* 49, 274-84.

Weber, A. (1909) *Uber den Standort der Industrien*; translated 1929 by C.J. Friedrich, Chicago, University of Chicago Press.

Weiss, S.T. and Gooding, E.C. (1970) 'Estimation of differential employment multipliers in a small region', in Richardson, H.W. (ed.) *Regional Economics: A Reader*, London, Macmillan.

West Central Scotland Planning Team (1974) *West Central Scotland Plan: Supplementary Report 1: The Regional Economy, West Central Scotland*, Glasgow.

Wheaton, W.C. (1977) 'A bid rent approach to housing demand', *Journal of Urban Economics* 4, 200-17.

Wilson, A.G. (1970a) *Entropy in Urban and Regional Modelling*, London, Pion.

Wilson, A.G. (1970b) 'Disaggregating elementary residential location models', *Papers of the Regional Science Association* 24, 103-25.

Wilson, A.G. (1972) 'Behavioural inputs to aggregative urban systems models', in Wilson A.G. (ed.), *Papers in Urban and Regional Analysis*, London, Pion.

Wilson, A.G. (1974) *Urban and Regional Models in Geography and Planning*, London, Wiley.

Wilson, A.G. (1981) *Geography and Environment*, Chichester, Wiley.

Wingo, L. (1961) *Transportation and Urban Land*, Washington, DC, Resources for the Future.

Wirth, L. (1938) 'Urbanism as a way of life', *American Journal of Sociology* 44, 3-24.

Wolpert, J. (1965) 'Behavioural aspects of the decision to migrate', *Regional Science Association Papers* 15, 159-69.

Wolpert, J., Mumphrey, A. and Seley, J. (1972) 'Metropolitan neighbourhoods: participation and conflict over change', Association of American Geographers, *Commission on College Geography, Resource Paper 16*, Washington, DC.

Zipf, G.K. (1949) *Human Behaviour and the Principle of Least Effort*, Reading, Mass., Addison-Wesley.

Name index

Ackoff, R.L., 12, 71
Alonso, W., 142, 144, 145, 149, 166
Althusser, L., 26, 26, 27, 36, 166
Archer, B.H., 52
Arrow, K.J., 33, 34, 122
Ashby, W.R., 42
Ashcroft, B., 52
Atkin, R.H., 83

Bacon, F., 17
Ball, M.J., 150, 166
Bannister, D., 114
Barr, A., 106
Barras, R., 61, 65, 71
Barrett, F.A., 147
Bassett, K., 165
Batty, M., 93
Baxter, R., 93
Bayes, T., 71, 108
BBC, 125
Beckhelm, T.L., 48
Beckmann, M.J., 144
Bell, W., 85
Bennett, R.J., 87
Bentham, J., 30
Berger, P.L., 18, 23
Bertalanffy, L., 142
Bhaskar, R., 23
Billings, R.B., 55
Boddy, M.J., 166
Bourdieu, P., 23
Bracken, I., 61, 65, 74
Breheny, M.J., 73

Broadbent, T.A., 26, 61, 65, 71
Brown, D.J., 47, 84, 149, 158
Brownrigg, M., 52
Burgess, E.W., 85
Buttimer, A., 21
Buttler, H.-J., 142

Camhis, M., 6
Cant, D.H., 166
Carrothers, G.A.P., 90
Castells, M., 26, 26, 148, 163, 166
Catanese, A.J., 24
Chadwick, G., 24
Chalmers, J.A., 48
Chapin, F.S., 125, 129, 132
Chisholm, M., 147
Cleveland, Middlesbrough, 57
Cockburn, C., 26, 170
Cole, R.L., 35
Comte, A., 16
Condorcet, 34
Coupe, R.T., 158
Cox, K.R., 186, 190
Cullen, I.G., 83, 107, 111, 113, 134, 137, 188
Curry, L., 87, 91, 93, 99

Dear, M., 26
Department of the Environment, 86, 155
Diamond, D.B., 150
Dicken, P., 157
Dickens, P., 169
Downs, R., 114

Dror, Y., 24
Durkheim, E., 189

Erickson, R.A., 157
Evans, A., 143, 150
Eversley, D., 12
Ewens, S., 35

Fagence, M., 113, 117
Feigenbaum, E.A., 106
Filmer, P., 18
Foot, D., 93
Ford, R.G., 147
Fothergill, S., 48
Friend, J.K., 25, 25, 27, 29

Galbraith, J.K., 154
Garnick, D.H., 55
Gatrell, A.C., 82, 83
Gershuny, J.I., 128
Giddens, A., 18, 23
Gilmour, J., 154
Goddard, J.B., 157
Godson, V., 134
Goldman, T., 144
Goldner, W., 92
Goldsmith, M., 35
Gooding, E.C., 52
Gould, P., 82, 83
Greater Manchester, 47
Gregory, D., 2, 3
Greig, M.A., 52
Gudgin, G., 48
Gupta, S.K., 70

Habermas, J., 18
Hagerstrand, T., 97, 129, 130, 132
Haggett, P., 81
Haimes, E.V., 134
Hall, P., 155
Hammond, S., 134
Harloe, M., 26
Harvey, D., 1, 2, 17, 21, 148, 149,
   159, 163, 166
Hegel, G.W.F., 189
Herbert, D.T., 85, 145, 153, 158
Hicks, J.R., 30
Hirschman, A.O., 44
Hoinville, G., 119, 122

Holtermann, S., 85
Hordijk, L., 87
Hotelling, H., 143
Houston, D., 47
Hoyt, H., 85
Hume, D., 21, 61, 65, 179

Ingram, G.K., 145
Institute of Operations Research,
   London, 25
Isard, W., 60, 144

Jencks, C., 32
Jessop, W.N., 25
Johnson, K.H., 87
Jones, P.M., 119, 122

Kain, J.F., 145, 150
Kaldor, N., 30
Kant, I., 29, 57
Keeble, D., 161
Keynes, J.M., 52
King, L.J., 81, 178
Kirwan, R.M., 150
Koopmans, T.C., 144
Kuenne, R.E., 144
Kuhn, H.W., 18, 20, 144

Lancashire, 161
Launhardt, W., 142
Lee, T., 114
Lenntorp, B., 97, 131
Leontieff, W., 52, 56
Ley, D., 190
Lichfield, N., 123
Little, I.M.D., 33, 193
London, 161, 162
Losch, A., 142, 143
Lowry, I., 92, 103, 104, 106
Luckmann, T., 18, 23
Lund, 129, 130
Lyon, H.L., 87

Mackay, D.I., 47
McKenzie, R.D., 85
McNicholl, I.H., 57
Mcloughlin, J.B., 24
Mair, J.M.M., 114
Marshak, T., 144
Martin, R.L., 87

Marx, K., 26, 26, 28, 31, 35, 38, 92, 162, 163, 165, 166, 168, 169, 170, 186, 190
Masser, I., 84
Massey, D.B., 148, 164, 169, 170
Meegan, R., 164, 169, 170
Merrett, A.J., 123
Michelson, W., 125, 128, 129, 188
Miernyk, W.H., 56
Milgram, S., 163
Mill, J.S., 30
Mishan, A.J., 30
Moore, B., 47, 48, 158
Morgan, B.S., 158
Morris, D.M., 157
Morrison, W., 57
Moser, C.A., 85
Moses, L., 56
Muth, R.F., 142, 144, 166
Myrdal, G., 44

NBER, 145
Newton, I., 87, 90
Nijkamp, P., 87
Northern Region, 47, 57

Oeppen, J.E., 87
Ohls, J.C. 165
Olsson, G., 90
O'Neill, J., 20
Openshaw, S., 84

Pareto, V., 30, 34
Parkes, D.N., 130
Perraton, J., 93
PESASP, 131
Phelps, E., 134
Plato, 113
Popper, K., 18, 20, 21, 89, 133
Poulantzas, N., 163, 166
Power, J.M., 25, 27
Pred, A., 3, 152
Pressat, R., 60

Quigley, J.M., 150

Randall, J.N., 47
Rawls, J., 30, 31, 32, 34
Rees, J., 63, 154
Rein, M., 113

Relph, E., 188
Rhodes, J., 47, 48
Richardson, H.W., 145, 150
Riefler, R., 56
Roberts, A.J., 73
Robinson, J.P., 125, 128, 129
Rosenhead, J., 70
Rossi, P., 158
Rousseau, J.-J., 29
Roweis, S.T., 6, 26, 113, 148, 166, 192
Rowley, G., 119

Salling, M., 159
Samuelson, P.A., 33, 49, 125
Sasieni, M.W., 71
Saunders, P., 35
Sawers, L., 26
Sayer, R.A., 2, 21, 92
Scott, A.J., 6, 26, 85, 113, 148, 166, 192
Scottish Development Agency, 161
Sen, A.K., 33
Senior, M.L. 145
Shetland, 57
Shevky, E., 85
Short, J.R., 158, 165
Simon, H.A., 133, 153, 159
Smith, D.M., 81, 85, 143, 144, 147, 150
Speare, A., 158
Stea, D., 114
Stegman, M.A., 147
Steiss, A.W., 24
Stevens, B.H., 144, 145
Storper, M., 148
Stringer, P., 35
Swales, J.K., 52
Sykes, A., 123
Szalai, A., 125, 128, 129

Tabb, W.K., 26
Thomas, G.S., 128
Thorngren, B., 152, 155, 156, 157
Thornley, A., 113
Thrift, N.J., 3, 130
Thunen, J.H. von, 142, 144, 149, 150
Tiebout, C.M., 52, 56

Tipple, G., 119
Tobler, W.R., 87
Tornquist, G., 154
Townroe, P., 152, 157, 162

Ullman, E.L., 85
United Kingdom, 25, 35, 57, 61,
    71, 85, 86, 128, 161, 164
United States, 128

Wales, 61
Walker, G., 57
Webber, R.J., 85
Weber, A., 142, 143, 144, 147, 149,
    150, 151
Weiss, S.T., 52
West Central Scotland, 47
West Midlands, 161, 162
Wheaton, W.C., 150
Wilson, A.G., 63, 90, 99, 119, 145
Wingo, L., 144
Wirth, L., 163
Wolpert, J., 35, 152, 158, 159, 190

Yewlett, C.J.L., 25, 27

Zipf, G.K., 87

Subject index

aggregate analysis: demographic, 58-9; descriptive, 69; economic, 43-5; projective, 70; spatial, 79

aggregate structural dynamics, 67-9

analysis: comparative static, 176-7; comparative statics description, 177; comparative statics prediction, 178-81; economic and demographic, 40-2; integrated, 71-3; integrated inc. state activity, 73-4; modes, 76-7, 174-5; projective, 42; relational dynamic, 181-3

analysis rules: deductive logic, 21; falsification principle, 21-2; inductive flexibility, 22-3; reflexive, 22

artificial intelligence, 106-7

behaviour analysis, 109-11; methodological myopia, 131-2

behavioural realism, 111

cohort survival analysis: applications, 65; assumptions and method, 62-4; limitations, 64-5

deductive logic, 21

deductive scientific method, 18

demographic analysis, 40-2; aggregate, 58-9

descriptive analysis, aggregate, 69

dialectical explanation, 188-90

dynamic analysis: competing models, 185-6; dialectical explanation, 188-90; relational, 181-3; sympathetic interpretation, 187-8; synthesis, 191

economic analysis, 40-2; aggregate, 43-5

export-base analysis: applications, 51-2; assumptions and method, 48-50; limitations, 50-1

factorial ecology, 85-6

falsification principle, 21-2

impossibility theorem, 34-5

inductive logic, 22-3

inductive scientific method, 17-18

input-output analysis: applications, 57; assumptions and method, 52-4; limitations, 54-7

integrated analysis, 71-3; inc. state activity, 73-4

integrated spatial analysis, 103-6

interaction analysis, 87-9; applications, 92; assumptions and method, 89-91; disaggregation, 99-101; limitations, 91-2; predictive, 94-5

lifestyle analysis, 125-6 integration, 132-7

location analysis: mechanical, 170-2; process, 139-42
location theory, 79

map pattern analysis, 80-1; applications, 83; assumptions and method, 81-2; limitations, 82-3
materialist analysis, 161-2; applications, 169-70; assumptions and method, 162-7; limitations, 167-9
materialist social choice, 31-2
'maximin' social choice, 30-1

negative feedback, 75-6
neoclassical analysis, 142-3; applications, 149-51; assumptions and method, 143-6; limitations, 146-9,

optimality criteria, 32-3

pattern analysis: factorial ecology, 85-6; predictive, 86-7; zoning, 84
perception analysis, 113; applications, 117; assumptions and method, 113-14; limitations, 114-17
planning: component activities, 14; decomposition theories, 15; generic, 11-13; holistic, 15-16; humanist reintegration, 194-200; investigative, 16-17; materialist perspective, 192-3
planning theory: humanist, 27, 36-8; process analysis, 27-8; rational, 24-5; structuralist, 26
projective analysis, 42; aggregate, 70
public participation, 33, 113; rates, 35

reflexive analysis, 22
revealed preference analysis: applications, 128; assumptions and method, 126-7; limitations, 127-8
risk analysis, 70-1

routinization, 137-9

scientific method: deductive, 18; inductive, 17-18; problem of interference, 19; problem of reduction, 19-20; problem of relativity, 20-1
scientific research, 18
shift and share analysis: applications, 47-8; assumptions and method, 45-6; limitations, 46-7
social change: alienated response, 185; human process, 184-5; pattern, 173-4
social choice: materialist, 31-2; 'maximin', 30-1; utilitarianism, 29-30
spatial analysis: aggregate, 79; integration, 103-6; utility, 101-3
spatial autocorrelation, 94
spatial form, 78-9; structural status, 97-8
static analysis, comparative, 176-7; descriptive, 177; predictive, 178-81
structural dynamics, aggregate, 67-9
sympathetic interpretation, 187-8
systems analysis, 152-3; applications, 161; assumptions and method, 153-9; limitations, 159-61

time geographic analysis: applications, 130-1; assumptions and method, 128-30; limitations, 130
trend analysis: applications, 61-2; assumptions and method, 59-60; limitations, 60-1

utilitarian social choice, 29-30

values analysis, 117-18; applications, 124; assumptions and method, 118-20; interdependence, 118; limitations, 120-4

zoning analysis, 84

For Product Safety Concerns and Information please contact our EU
representative GPSR@taylorandfrancis.com Taylor & Francis Verlag GmbH,
Kaufingerstraße 24, 80331 München, Germany

Printed and bound by CPI Group (UK) Ltd, Croydon, CR0 4YY
01/05/2025
01858581-0001